Lecture Notes in Mathematics

Edited by A. Dold and B. Eckmann

1145

Gerhard Winkler

Choquet Order and Simplices

with Applications in Probabilistic Models

Springer-Verlag
Berlin Heidelberg New York Tokyo

Author

Gerhard Winkler
Mathematisches Institut, Universität München
Theresienstr. 39, 8000 München 2, Federal Republic of Germany

Mathematics Subject Classification (1980): primary: 46A55
secondary: 18B99, 28C99, 46E27, 52A07, 60G05

ISBN 3-540-15683-6 Springer-Verlag Berlin Heidelberg New York Tokyo
ISBN 0-387-15683-6 Springer-Verlag New York Heidelberg Berlin Tokyo

Printing and binding: Beltz Offsetdruck, Hemsbach/Bergstr.
2146/3140-543210

PREFACE

These lecture notes present a summary of certain properties of Choquet order on locally convex spaces, an examination of inverse systems of simplices and applications in probability theory. The central results were obtained by the author in 1982 and 1983. There is little or no overlap with recent surveys - like R.D. Bourgin's monograph (1983).

The reader I had in mind when writing down the *proofs* is a student of mathematics attending a seminar in the last third of his study. So the proofs are rather thorough and the specialist will probably skip a lot of details. The reader is assumed to be familiar with the basic ideas of topological measure theory and locally convex vector spaces. It is useful but not necessary to have some familiarity with the Choquet theory.

I am indebted to J.P.R. Christensen, S. Dierolf, G. Godefroy, Chr. Hele, H. Kellerer, E. Kolb, R. Kotecký, Z. Lipecky, D. Preiss and last but not least H.v. Weizsäcker for their help and useful comments.

CONTENTS

INTRODUCTION

Choquet order is one of the basic tools in the theory of integral representation. Consider a convex set M in some locally convex linear space E. An element x of M is the barycenter of a probability measure p on M - or it is represented by p - iff

$$l(x) = \int_M l(y)\, dp(y) \quad \text{for every continuous linear functional } l \text{ on } E.$$

One asks for those sets M where every element has such a representation with a measure p living on the extreme points of M. In its different equivalent formulations, Choquet order served G. Choquet, G.A. Edgar and many others to answer this question on various levels of generality. The theory for compact convex sets is surveyed in the nowadays almost classical books of R.R. Phelps (1966), G. Choquet (1969) and E. Alfsen (1971). L. Asimov and A.J. Ellis (1980) describe later development. Since the monographs of R.D. Bourgin (1983) and B. Fuchssteiner and W. Lusky (1981) which cover the noncompact situation, presently there remains not too much to say about this topic. So we do not pursue this aspect of Choquet order. It only appears in the remarks rounding off chapter 1.

The emphasis of the present notes is different. We start studying Choquet order in its own right, then investigate systems of simplices and try finally to bridge the gap to applications, which we choose according to our personal taste from probability theory. To avoid restrictions in applicability, we have to work in a fairly general setting which is also of interest in the theory itself.

To be more specific, I will comment on the single chapters below.

As we will speak only about probability measures on a locally convex space which are inner regular w.r.t. compact sets, let us simply call them "measures" in this introduction. Choquet order measures how far "outside" the mass of some measure p is distributed compared with some other measure q. So both measures are tested by means of convex functions. Write

$$p < q \quad \text{iff} \quad \int \varphi \, dp \leq \int \varphi \, dq \quad \text{for every convex continuous function } \varphi,$$

provided the integrals exist and call p smaller than q in Choquet order.

The first two chapters are devoted to Choquet order itself. Chapter 1 is a collection of basic properties. This requires a lot of space but it fills at least partially a gap in the presently available literature. Many of the important facts are scattered in various papers, sometimes only hidden in the proofs, and some are not even published.

First, we introduce barycenters and prove R. Haydon's theorem: barycenters exist in the completion of the underlying locally convex space and depend continuously on the measures. Next, we have to decide on which kind of sets the measures are supposed to live. On the one hand, the class of such sets should be large enough to cover those considered in Choquet theory up to now as well as the applications we have in mind. On the other hand pathological behaviour should be excluded. It turns out to be sufficient to require that each measure on such a set have a barycenter and that this barycenter do not fall out of the set. This property is a weak version of completeness, called measure convexity. It guarantees for instance, that the Dirac measure of an extreme point is as one would expect maximal in Choquet order. As soon as this problem is settled, equivalent formulations of Choquet order (on arbitrary sets) are collected: definitions by means of cones of functions, the dilation order already used by Choquet, the description by means of conditional expectations which is the link between Choquet order and martingales and a new characterization due to H.v. Weizsäcker. The latter is a converse to the dilation order. We then study boundary measures - those which are maximal in Choquet order and which "have their mass as close as possible near the extreme points"-and introduce simplices. In contrast to Choquet's lattice theoretic definition, we prefer the one most natural in the present context: a subset S of a locally convex space is a simplex iff each of its elements is the barycenter of one and only one boundary measure (and - of course - if it is measure convex). Several useful properties are derived and the relation with Choquet simplices is examined. So far, nothing is really new.

New results concerning Choquet order are presented in chapter 2. Naively, one would expect that measures smaller in Choquet order live on smaller sets. Fortunately, this is true if the sets are measure convex (but fails as soon as they are not). Measures on measure convex sets have a strong tightness property: the set can be exhausted by convex compact sets in measure. In the second section, we see that this has the consequence that sets of measures which are bounded above (by a measure concentrated on a measure convex set) are uniformly tight w.r.t. convex compact sets. Subsequently, the extension problem is attacked: when has a measure on the Borel σ-algebra of the weak topology an extension to the strong one? A close connection with Choquet order is revealed. We also reprove the classical result for metrizable locally convex spaces. In section 2.4, we study decreasing nets in Choquet order (increasing nets have been examined exhaustively in connection with existence of representing measures). On measure convex sets they have a greatest lower bound and converge to it weakly. This result will be the basis of the investigations in chapter 3.

By way of introduction to chapter 3, let me pose an elementary problem.

Consider a decreasing sequence $S_1 \supset S_2 \supset S_3 \supset \dots$ of simplices in Euclidean n-space (by a simplex I now mean a point, compact line-segment, triangle,...). Is their intersection a simplex? Not surprisingly, the answer is "yes" (a translation of Borovikov's elementary solution (1952) is included). The answer is also "yes" for decreasing nets of infinite dimensional compact Choquet simplices. More generally, the inverse limit of an inverse system of compact Choquet simplices is again a compact Choquet simplex as E.B. Davies and G.F. Vincent-Smith (1968) could prove. In section 3.3, we review the historical development. However as often happens in Choquet theory, compactness is an obstacle for application: it may be cumbersome or even impossible to construct artificially reasonable topologies in which the sets in question are compact. This nuisance is remedied by showing that the inverse limit of noncompact simplices is a simplex in all reasonable situations. The proof is based on the properties of Choquet order derived in chapter 2. The results cannot be improved, as counterexamples show.

The structures studied in chapter 3 are inherent in models from various fields of mathematics. In chapter 4, they are by way of example applied to a problem which at first glance looks purely measure-theoretic. The following is a consequence of the results about inverse limits of simplices (and it proves to be convenient that we worked with noncompact simplices): inverse limits exist in the category of standard Borel spaces and substochastic kernels. This may be interpreted in concrete models from probability theory, statistics and statistical mechanics; e.g. for Markov processes it means the construction of Dynkin's entrance boundary. Further examples are sketched in passing; the geometrical structure of sets of Gibbs states is treated in more detail.

One may argue that a purely measure-theoretic problem should be solved by purely measure-theoretic methods. But the view taken here allows a very natural and transparent proof and a better understanding of the problem. D. Preiss' derivation of a well-known criterion for existence of Gibbs states is another example in the same spirit (cf. section 4.2 and 4.3.2). In summary, I wanted to show that it can be pleasant and useful to look at probabilistic objects through geometrical spectacles.

CHAPTER 0

NOTATIONS, DEFINITIONS AND CONVENTIONS

Notations, definitions and conventions used throughout the paper are collected in this chapter. The reader who skips it and starts with chapter 1 should notice the conventions in 0.3, 0.6 and 0.7 typed in italics.

0.1 We denote the set of positive natural numbers by $\mathbb{N}$, the set of integers by $\mathbb{Z}$, the set of real numbers by $\mathbb{R}$, the set of nonnegative real numbers by $\mathbb{R}_+$ and Euclidean n-space by $\mathbb{R}^n$.

0.2 By a measure on a measurable space (X,F) we always mean a nonnegative finite σ-additive set function (with real values) on the σ-algebra F. A measure p is called a probability measure iff $p(X) = 1$. A signed measure is a set function of the form $q = p_1 - p_2$ where p_1 and p_2 are measures.

Denote by

$B(X,F)$	the space of real bounded measurable functions on X,
$P(X,F)$	the set of probability measures on F,
$M_+(X,F)$	the positive cone of measures on F - including the zero-measure,
$M(X,F)$	the space of signed measures on F.

The integral of a measurable function f on X w.r.t. $p \in M(X,F)$ is denoted by $\int_X f(x)\,dp(x)$, $\int_X f\,dp$ or simply $\int f\,dp$. If f is bounded, we abbreviate it by $p(f)$. For $A \subset X$ the inner measure is

$$p_*(A) := \sup\{p(B): B \in F, B \subset A\},$$

the outer measure

$$p^*(A) := p(X) - p_*(X \setminus A);$$

the <u>p-completion</u> of F is

$$F_p := \{A \subset X: p_*(A) = p^*(A)\}.$$

The unique extension of p to F_p is again denoted by p.

If $X_o \subset X$ and $p \in M_+(X_o, X_o \cap F)$, then we often identify the canonical extension of p to F with p without further remark.

If $x \in X$, the <u>point measure</u> ε_x is defined by $\varepsilon_x(B) := 1_B(x)$ for every $B \in F$.

Let (X,F) and (Y,G) be measurable spaces. A mapping

$$P : X \to M(Y,G),\ x \to P(x,\cdot)$$

is called a <u>(sub-)stochastic kernel</u> from X to Y whenever

(i) $P(x,\cdot) \in M_+(Y,G)$ for every $x \in X$,

(ii) $P(x,Y) = 1$ $(P(x,Y) \leq 1)$ for every $x \in X$,

(iii) the mappings $P(\cdot,B)$ are measurable for each $B \in G$.

If Q is a substochastic kernel from Y to some measurable space (Z,H), the composition PQ is a substochastic kernel from X to Z defined by

$$PQ(x,B) := \int_Y Q(y,B)\, dP(x,y) \text{ for every } B \in H.$$

P induces a linear mapping from $M(X,F)$ to $M(Y,G)$ via

$$M(X,F) \ni \mu \to \mu P \in M(Y,G),$$

$$\mu P(B) := \int_X P(x,B)\, d\mu(x) \text{ for every } B \in G$$

and a linear mapping from $B(Y,G)$ to $B(X,F)$ via

$$B(Y,G) \ni f \to Pf \in B(X,F),$$

$$Pf(x) := P(x,f) := \int_Y f(y)\, dP(x,y) \text{ for every } x \in X.$$

0.3 *Topological spaces are assumed to be separated throughout the paper.*

The use of terms "Suslin" and "Lusin" spaces is not unique in literature.

Definition. Let X be a topological space.

a. X is said to be a Polish space iff it is separable and there is a metric on X compatible with the topology in which X is complete.

b. X is called Lusin space iff there is a Polish space Y and a continuous bijective mapping from Y onto X.

c. X is said to be a Suslin space iff there is a Polish space Y and a continuous surjection from Y onto X.

Remark. Lusin spaces are often called "standard" and Suslin spaces are also called "analytic" spaces, e.g. in Hoffmann-Jørgensen (1970).

Now, we can define an important measure theoretic notion.

Definition. A measurable space (X,F) is called a standard Borel space iff it is Borel isomorphic to a Polish space (Y,t), i.e. there is a bijection $\varphi : Y \to X$ such that φ as well as φ^{-1} are measurable w.r.t. F and the σ-algebra generated by t.

Remark. We have given the definition most common in literature, using the topological notion of a Polish space. A purely measure theoretic characterization can be given as well. In fact, Dynkin (1978) uses only notions from measure theory to define "B-spaces" which coincide with standard Borel spaces; cf. chapter 4.3.4.

If X is topologized by t, then we denote by

$C(X,t)$ the space of real continuous functions on X,

$C_b(X,t)$ the space of bounded elements from $C(X,t)$,

$\mathcal{B}_o(X,t)$ the Baire-σ-algebra, generated by $C_b(X,t)$,

$\mathbf{B}(X,t)$ the Borel σ-algebra, i.e. the σ-algebra generated by t,

$\mathcal{B}(X,t)$ the space $\mathcal{B}(X,\mathbf{B}(X,t))$.

If (X,s) and (Y,t) are topological spaces, the product topology on $X \times Y$ is denoted by $s \times t$. If there are topological spaces (X_i,t_i), $i \in I$, the product topology on $\prod_{i\in I} X_i$ is denoted by $\prod_{i\in I} t_i$.

0.4 Let X be a topological space with topology t. A measure p on $\mathbf{B}(X,t)$ is called τ-smooth iff for every decreasing net $(A_i)_{i\in I}$ of closed sets we have

$$p\left(\bigcap_{i\in I} A_i\right) = \inf\{p(A_i) : i \in I\}.$$

It is straightforward to show for a τ-smooth measure p: Let $(f_i)_{i\in I}$ be an increasing net of lower semi-continuous functions on X, integrable w.r.t. p and denote by f its supremum. Then:

$$\int f \, dp = \sup\left\{ \int f_i \, dp : i \in I \right\}$$

(the integral may attain the value $+\infty$).

For a τ-smooth measure p there is a smallest closed set of full measure - called support of p.

A measure p on $\mathbf{B}(X,t)$ is called tight iff

$$p(B) = \sup\{p(C) : C \subset B,\ C \text{ compact}\} \text{ for every } B \in \mathbf{B}(X,t).$$

Every tight measure is τ-smooth.

We abbreviate by

$P(X,t)$ the set of tight probability measures,

$M_+(X,t)$ the positive cone of tight measures - including the zero-measure ,

$M(X,t)$ the linear space $M_+(X,t) - M_+(X,t)$.

A subset A of X is called universally measurable iff it is in the p-completion of $\mathcal{B}(X,t)$ for every tight measure p on X.

0.5 If there is no danger of confusion, we suppress the topology or σ-algebra and write $C(X)$, $P(X)$, $M(X)$ and so forth.

0.6 *Only vector spaces over the real field will be considered.*

A vector x is the convex combination of vectors $x_1,\dots,x_n$ iff there are $a_1,\dots,a_n \in \mathbb{R}_+$ with $\sum_{i=1}^{n} a_i = 1$ such that $x = \sum_{i=1}^{n} a_i x_i$. A subset C of a vector space is convex iff it contains all convex combinations of its elements. An element x of a convex set C is extremal or an extreme point iff $C \setminus \{x\}$ is convex; the collection of extreme points is called extreme boundary and denoted by ex C. If M is a subset of a linear space H then the convex hull co M of M is the smallest convex subset of H containing M. A subset C of H is a cone iff $\mathbb{R}_+ \cdot C \subset C$ and a convex cone iff it is in addition convex. A real function f on a convex set C is said to be convex iff

$$f(ax + (1-a)y) \leq af(x) + (1-a)f(y)$$

whenever $x,y \in C$ and $0 < a < 1$; it is strictly convex iff always the strict inequality holds and affine iff always equality holds.

H* denotes the algebraic dual space of H.

0.7 (E,τ) denotes a (real, separated!) locally convex space, E' its topological dual space. Subsets always have the relative topology.

Let M be a subset of E. The closed convex hull $\overline{\text{co}}$ M of M is the smallest closed convex subset of E containing M.

$Z(M)$ is the σ-algebra of "cylinder sets" on M, generated by E',

$S(M)$ denotes the cone of all functionals $\varphi : M \to \mathbb{R}$ of the form

$\varphi = \max_{i \leq n} (l_i(\cdot) + c_i)$ where $l_i \in E'$ and $c_i \in \mathbb{R}$, $1 \leq i \leq n$, for some $n \in \mathbb{N}$.

Two properties of the cone $S(M)$ will be needed.

Lemma. Let M be a subset of the locally convex space E. Then:

a. The vector space $S(M) - S(M)$ spanned by $S(M)$ is a vector lattice in the pointwise order.

b. Suppose that $p, q \in P(M)$ such that every $\varphi \in S(M)$ is integrable w.r.t. p and q. Then $\int \varphi \, dp = \int \varphi \, dq$ for every $\varphi \in S(M)$ implies $p = q$.

Proof. Obviously $S(M)$ contains with two functions φ and ψ also the pointwise maximum $\varphi \vee \psi$. Moreover, $|\varphi - \psi| = 2(\varphi \vee \psi) - (\varphi + \psi)$; this shows that $S(M) - S(M)$ contains the absolute value of any of its elements, hence is a vector lattice which proves a.

Let us now prove b. The space $L := S(M) - S(M)$ is a vector lattice by a. and contains the constants. So by the uniqueness assertion of the Daniell extension theorem, p is equal to q on the σ-algebra generated on M by L, that is $Z(M)$. Choose now compact sets C_n, $n \in \mathbb{N}$, such that for their union C

$$p(C) = 1 = q(C).$$

This choice is possible, since p and q are tight. We show now that

$$\mathcal{B}_o(C) = Z(C).$$

In fact, every $f \in C_b(C)$ is the pointwise limit of a sequence from L, hence $Z(M)$-measurable: for every $n \in \mathbb{N}$ there is by the lattice version of the Stone-Weierstraß theorem some $\varphi_n \in L$ with

$$\sup_{x \in C_n} |f(x) - \varphi_n(x)| < \frac{1}{n}.$$

By uniqueness of the extension of tight Baire measures, p and q agree on $\mathcal{B}(C)$. This completes the proof of b. □

If H is a linear space and G a subspace of H^* then $\sigma(H,G)$ is the topology on H generated by the mappings $x \to g(x)$, $g \in G$.

If (X,t) *is a completely regular topological space, then* $M(X)$ *is always endowed with the weak topology* $\sigma(M(X),C_b(X))$
generated by the functionals $\mu \to \mu(f)$, $f \in C_b(X)$, which makes it a locally convex space. Convergence in this topology will be indicated by the symbol $\mu_i \overset{w}{\to} \mu$.

0.8 As standard references, we use Dugundji(1968) for general topology, Hoffmann-Jørgensen(1970) and Schwartz(1973) for the theory of Suslin and Lusin spaces, Topsøe(1973) and also Schwartz(1973) for topological measure theory and Schaefer(1973) for topological vector spaces.

CHAPTER 1

BASIC CONCEPTS FROM NONCOMPACT CHOQUET THEORY

There are several excellent surveys for special fields in Choquet theory. Some covering the case of compact convex sets may meanwhile be called classical - like Phelps(1966) and Alfsen(1971). Bourgin(1983) describes the theory for bounded closed convex subsets of Banach spaces. We will work in a more general framework, which will be established in the following sections. It is appropriate for the investigations in chapter 2 and 3; in chapter 4 we will show that it is useful for applications. If one is interested in such a general setting, the search for a standard reference is idle. It is beyond the scope of the present paper to provide a systematic and exhaustive treatment of the foundations. So I had to confine myself to put together at least those basic concepts which are indispensable to understand the subsequent studies.

Nothing in this chapter is really new. Nevertheless, the proofs are elaborated. The reason is, that we need results from v. Weizsäcker(1977) which are not published in current mathematical journals; also a lot of arguments is scattered and hidden in various papers. Several remarks indicate the connection with related material, sketch the historical background and contain suggestions for further reading.

Section 1.1. Barycenters, representing measures and the barycenter map

Let us start with the basic notion of a barycenter. Intuitively, one may think of the centre of mass of some material distributed in space.

1.1.1 Definition. A point x in a locally convex space E is called the barycenter of the probability measure $p \in \mathcal{P}(E)$ iff

(i) each $l \in E'$ is integrable w.r.t. p,

(ii) $l(x) = \int l\, dp$ for every $l \in E'$.

We also say that p represents x and denote x by $r(p)$. The expression in (ii) will be called barycentrical formula and the map $p \to r(p)$ the barycenter map (provided it is defined).

Note that barycenters are defined in terms of the Pettis integral and if they exist, they are unique since E' separates points.

Remark. If M is bounded then condition (i) is fulfilled.

We note three simple special cases.

Example 1. If the mass is concentrated in one point x - i.e. $p = \varepsilon_x$ - then the barycenter is x itself. More generally if $p = \sum_{i \leq n} a_i \varepsilon_{x_i}$ is a convex combination of point masses, then $r(p) = \sum_{i \leq n} a_i x_i$.

Example 2. Similarly, let $p = \sum_{i=1}^{\infty} a_i \varepsilon_{x_i}$, where $x_i \in E$, $a_i \geq 0$, $\sum_{i=1}^{\infty} a_i = 1$; then x is the barycenter of p iff

(i) $\sum_{i=1}^{\infty} a_i \mid l(x_i) \mid < \infty$ for every $l \in E'$,

(ii) $l(x) = \sum_{i=1}^{\infty} a_i l(x_i)$ for every $l \in E'$.

In this case, $x = \sum_{i=1}^{\infty} a_i x_i$ in the weak topology $\sigma(E,E')$.

Example 3. a. Consider some completely regular space X and let $E := M(X)$. Recall that $M(X)$ is endowed with the topology $\sigma(M(X), C_b(X))$. If $\mu \in M(X)$ is the barycenter of $p \in P(M(X))$ then condition (ii) in definition 1.1.1 reduces to

$$\mu(f) = \int \nu(f)\, dp(\nu) \quad \text{for every } f \in C_b(X).$$

In proposition 1.1.2 we will see, that this formula holds also for more general functions f.

b. It is easy to see that ex $P(X) = \{\varepsilon_x : x \in X\}$ and that this set is homeomorphic to X itself via the natural injection $\iota : x \to \varepsilon_x$ (Topsøe (1970), thm. 11.1). Choose now $\mu \in P(X)$; lift μ to the measure $\mu\circ\iota^{-1}$ on the extreme points of $P(X)$. The integral transformation formula yields

$$\mu(f) = \int_{\text{ex } P(X)} \nu(f)\, d\mu\circ\iota^{-1}(\nu)$$

such that μ is the barycenter of the measure $\mu\circ\iota^{-1}$ on ex $P(X)$. Further, it is clear that each $p \in P(\text{ex } P(X))$ has barycenter $p\circ\iota \in P(X)$. Hence the barycenter map is a bijection from $P(\text{ex } P(X))$ onto $P(X)$.

In applications, the barycentrical formula may not be adequate in the form 1.1.1(ii). For example, if one deals with measures, the barycentrical formula in example 3 should not only hold for bounded continuous functions, but also for measurable functions. As these lecture notes are written with a side-glance on probability theory, we confine ourselves to show how such a gap can be filled in this special context.

<u>1.1.2</u> <u>Proposition</u>. Let X be a completely regular space and $g : X \to \mathbb{R}$ a Borel measurable function. Suppose further that H is a subset of $\{\nu \in M_+(X) : g \in L^1(\nu)\}$. Then the evaluation map

$$\Phi : H \to \mathbb{R}, \ \nu \to \int g\, d\nu$$

satisfies for every $p \in P(H)$ with barycenter μ in H the barycentrical formula

$$\Phi(\mu) = \int_H \Phi\, dp.$$

In particular, setting $g = 1_B$ one gets

$$\mu(B) = \int_H \nu(B)\, dp(\nu) \quad \text{for each} \quad B \in \mathcal{B}(X).$$

Conversely, if the latter holds then μ is the barycenter of p.

Proof. From the definition of the topology on $M(X)$ we have

$$\mu(f) = \int_H \nu(f)\, dp(\nu) \quad \text{for every } f \in C_b(X).$$

Let U be an open subset of X and set $C := \{f \in C_b(X) : 0 \leq f \leq 1_U\}$. Then $\nu(U) = \sup\{\nu(f) : f \in C\}$ for each $\nu \in H$, since each ν is τ-smooth (cf. 0.4). μ and p are τ-smooth for the same reason, hence

$$\mu(U) = \int_H \nu(U)\, dp(\nu) \quad \text{for every open subset } U \text{ in } X.$$

By standard monotone class arguments the barycentrical formula holds for any $h \in \mathcal{B}(X)$ and by monotone convergence for $g^+ = g \vee 0$ and $g^- = -(g \wedge 0)$, hence for g itself. The rest is standard. □

The partial success achieved in the preceding proposition might support the belief that every (universally) measurable affine function Φ satisfies the barycentrical formula. This fails to be true.

Example 4. This example is due to Choquet(1962). Set $H := P([0,1])$ and define a function Φ on H by $\Phi(\nu) := \nu_s([0,1])$ where ν_s is the singular part of ν. Φ is obviously bounded and affine - moreover it is Borel measurable (for the proof the reader is referred to Alfsen(1971), p. 21). The point measures ε_x coincide with the extreme points of H (cf. ex. 3), consequently $\Phi|\,\mathrm{ex}\, H = 1$. On the other hand, for the Lebesgue measure λ one has $\Phi(\lambda) = 0$. In example 3, we have seen that $\lambda = r(p)$ for some p concentrated on $\{\varepsilon_x : x \in [0,1]\}$, hence

$$\Phi(r(p)) = \Phi(\lambda) = 0 \neq 1 = p(\Phi).$$

Especially the second part of the following proposition will be needed. The proof is due to R. Haydon (1976).

1.1.3 Proposition. Suppose that the locally convex space E is embedded in its completion $\tilde{E}$ and that M is a bounded subset of E. Then:

a. for every $p \in P(M)$ the barycenter $r(p)$ exists in $\tilde{E}$,

b. the barycenter map

$$r : P(M) \to \tilde{E}, \quad p \to r(p)$$

is affine and continuous.

Remark. Note that the barycenter map is not only continuous in the weak topology on $\tilde{E}$ - which is trivially true - but even in the original topology.

Proof of proposition 1.1.3.

1. Let us first reduce the problem to the case of Banach spaces. Suppose that 1.1.3 holds for Banach spaces. Since $\tilde{E}$ is complete it is isomorphic to an inverse limit of a family $(E_i)_{i \in I}$ of Banach spaces (Schaefer(1971), II.5.4). Assume that $\tilde{E}$ itself is this inverse limit and denote the canonical projection of $\tilde{E}$ into E_i by pr_i. Denote by x_i the barycenter of $p \circ pr_i^{-1}$, which exists by assumption. Assume now additionally, that each space $pr_i[\tilde{E}]$ is dense in E_i. This is no restriction, since we may take the closure of $pr_i[\tilde{E}]$ in E_i instead of E_i. $\tilde{E}$ is then called a reduced inverse limit. Proposition III.4.4 from Schaefer(1971) tells us that $\tilde{E}'$ is (algebraically) isomorphic to a direct limit of the spaces E_i' and it is straightforward to check that $x = (x_i)_{i \in I}$ is the barycenter of p in $\tilde{E}$. Since the topology on $\tilde{E}$ is the projective topology w.r.t. the family $(E_i, pr_i)_{i \in I}$, the barycenter map $p \to r(p)$ is continuous if and only if the maps $p \to pr_i(r(p))$ are continuous for every $i \in I$. But such a map is the composition of $p \to p \circ pr_i^{-1}$ and the barycenter map $r_i : P(pr_i[M]) \to E_i$ which is continuous by assumption.

2. Consider now a Banach space E. We may choose M to be the unit ball B of E. Denote by $\|\cdot\|$ the norm of E" and suppose that E is embedded into E". For each $p \in P(B)$ the element $r(p) \in E''$ is defined by $l \to p(l)$, $l \in E'$. We show first that

$$r : \mathcal{P}(B) \to E'' , \; p \to r(p)$$

is continuous w.r.t. $\sigma(\mathcal{M}(B), C_b(B))$ and $\|\cdot\|$. Choose $\varepsilon > 0$ and consider a covering $(U_\alpha)_{\alpha\in A}$ of B by open sets of diameter ε. A metrizable set is paracompact by Stone's theorem (Dugundji(1968), IX.5.3), so there is a partition of unity $(\kappa_\alpha)_{\alpha\in A}$ subordinated to $(U_\alpha)_{\alpha\in A}$ (Dugundji(1968), VIII.4). Recall that this means

(i) $\kappa_\alpha : B \to [0,1]$ is continuous,

(ii) the supports of the κ_α form a neighbourhood-finite covering of B,

(iii) $\sum_{\alpha\in A} \kappa_\alpha(x) = 1$ for each $x \in B$ (the sum is well-defined since $\kappa_\alpha(x) = 0$ for all but finitely many α),

(iv) the support of κ_α is contained in U_α.

Select for each α an element x_α in the support of κ_α and define the continuous mapping

$$f : B \to E, \; x \to \sum_{\alpha\in A} \kappa_\alpha(x)\, x_\alpha$$

(each f(x) is a convex combination of a finite collection of x_α's). By construction

(1) $\| f(x) - x \| \leqq \varepsilon$ for each $x \in B$.

Consider a net $(p_i)_{i\in I}$ converging weakly in $\mathcal{P}(B)$ to some measure p. Since every tight measure is τ-smooth, we have

(*) $p(\sum_{\alpha\in F} \kappa_\alpha) \uparrow 1$

where F runs through the finite subsets of A. Choose such an F with

(2) $p(\sum_{\alpha\notin F} \kappa_\alpha) < \varepsilon$.

Choose further $j \in I$ such that

(3) $$p_i(\sum_{\alpha \notin F} \kappa_\alpha) < \varepsilon \quad \text{for every } i \geq j,$$

(4) $$|p(\kappa_\alpha) - p_i(\kappa_\alpha)| < (\text{card } F)^{-1} \varepsilon \quad \text{for each } \alpha \in F \text{ and } i \geq j.$$

Write y for r(p) and y_i for $r(p_i)$ as defined at the beginning of this proof. The quantity to be estimated is

(5) $$\| y - y_i \| = \sup\{|l(y) - l(y_i)| : l \text{ in the unit ball of } E'\}.$$

For each l in the unit ball of E' we have by (1) that

(6) $$|l(y) - p(l \circ f)| \leq \varepsilon \quad \text{and} \quad |l(y_i) - p_i(l \circ f)| \leq \varepsilon$$

and

$$|p(l \circ f) - p_i(l \circ f)| = |\int l \circ f \, dp - \int l \circ f \, dp_i| \leq$$

$$\leq |\sum_{\alpha \in F} l(x_\alpha) \left[\int \kappa_\alpha \, dp - \int \kappa_\alpha \, dp_i \right]| +$$

$$+ |\int \sum_{\alpha \notin F} l(x_\alpha) \kappa_\alpha \, dp| + |\int \sum_{\alpha \notin F} l(x_\alpha) \kappa_\alpha \, dp_i| \leq$$

$$\leq \sum_{\alpha \in F} |p(\kappa_\alpha) - p_i(\kappa_\alpha)| + p(\sum_{\alpha \notin F} \kappa_\alpha) + p_i(\sum_{\alpha \notin F} \kappa_\alpha).$$

Applying (4), (2) and (3) to the last line we get

(7) $$|p(l \circ f) - p_i(l \circ f)| \leq 3\varepsilon \quad \text{for every } i \geq j.$$

Because of (5), a combination of (6) and (7) yields

$$\| r(p) - r(p_i) \| \leq 5\varepsilon \quad \text{for every } i \geq j,$$

hence the desired continuity of $r : P(B) \to E''$.

It remains to show that r maps P(B) actually into E. Because $r(\varepsilon_x) = x \in E$ for $x \in B$, r maps each convex combination of such point measures into E. Since they are dense in P(B) and r is continuous, r maps P(B) into the closure of E which is - due to completeness - E itself. □

Remark. R. Haydon (1976) shows continuity of the integral for every bounded continuous function on a completely regular space with values in a complete locally convex space on a broader class of measures which is defined by property (*). All his essential ideas appear in the above proof; to cover the full generality of Haydon's theorem requires additional notational efforts.

In proposition 1.1.3 completeness is essential. This is exhibited by the following counterexample in the setting of measures due to L. Schwartz ((1973a), (8,1) on page 145).

Example 5. There is a subset Y of the unit square $X := [0,1] \times [0,1]$ which is not measurable for the Lebesgue measure λ^2 on X and whose complement intersects each vertical line $\{x\} \times [0,1]$ in exactly one point (Gelbaum/Olmsted(1965), 10.23). Set $M := \mathcal{P}(Y)$; we construct $p \in \mathcal{P}(M)$ with no barycenter in $\mathcal{M}(Y)$.
Every measure of the form $\varepsilon_x \otimes \lambda$ is concentrated on Y and defines a tight measure λ_x on Y. Denote by θ the continuous map $[0,1] \ni x \to \lambda_x$ and set $p := \lambda \circ \theta^{-1}$, which is a tight probability measure on M. Define a non-negative σ-additive and normalized set function μ on $\mathcal{B}(Y)$ by

$$\mu(B) := \int_M \nu(B)\, dp(\nu) = \int_{[0,1]} \lambda_x\, d\lambda(x)$$

for every $B \in \mathcal{B}(Y)$.

Assume that p has a barycenter in $\mathcal{M}(Y)$. Then it must be μ by proposition 1.1.2. Each $B \in \mathcal{B}(Y)$ is of the form $B' \cap Y$ with some $B' \in \mathcal{B}(X)$ and

$$\lambda^2(B') = \int_{[0,1]} \varepsilon_x \otimes \lambda(B')\, d\lambda(x) = \int_{[0,1]} \lambda_x\, d\lambda(x) = \mu(B).$$

If μ were tight, it would be concentrated on a σ-compact subset C of Y and the preceding formula would imply $\lambda^2(C) = 1$. Then Y were λ^2-measurable in contradiction to the choice of Y. So $\mu \notin \mathcal{M}(Y)$.

Section 1.2. Measure convex sets

Measure convex sets are introduced. They form the natural setting for our investigations as will be justified in this and the following sections. Their class is small enough to avoid pathologies and large enough to cover the sets considered in non-compact Choquet theory in the past.

1.2.1 Definition. A subset M of a locally convex space is called measure convex iff for every $p \in P(M)$ the barycenter $r(p)$ exists and is contained in M.

Measure convex sets are of course convex.

Remark. Do not mix up measure convex sets with the "barycentrically closed" sets as defined in Fremlin/Pryce(1974), 2I. The latter contain each point which is the barycenter of some $p \in P(M)$. But in contrast to measure convex sets not every such measure must have a barycenter.

The first example shows that sets with nice measure theoretical and topological properties may fail to be measure convex.

Example 1. There is a bounded convex σ-compact (hence Borel measurable) Lusin subset M of a separable Banach space and $p \in P(M)$ with $r(p) \notin M$. Denote by e_i, $i \in \mathbb{N}$, the unit vectors of $\ell^1(\mathbb{N})$. Define $C_k := \mathrm{co}\,\{e_1,\ldots,e_k\}$ and $C := \bigcup_{k\geq 1} C_k$. The set C is σ-compact, in particular a Borel set. Since in the Polish space $\ell^1(\mathbb{N})$ each Borel set is a Lusin space (Schwartz(1973), thm. 2 on p. 95), C is a bounded convex Lusin subset of $\ell^1(\mathbb{N})$. The measure $p := \sum_{n\geq 1} 2^{-i}\varepsilon_{e_i}$ is in $P(C)$ but C does not contain its barycenter $x = \sum_{n\geq 1} 2^{-i} e_i$ (for $x = r(p)$ cf. the remark after definition 1.1.1 and example 2 in section 1.1). Obviously C is not complete (cf. corollary 1.2.4 below).

Before we give further examples and counterexamples, let us show that the members of an important class of sets *are* measure convex (corollary 1.2.4). We need a result due to D.H. Fremlin and J.D. Pryce ((1974), thm. 2E).

1.2.2 Proposition. Let M be a bounded convex subset of the locally convex space E. Suppose that $p \in P(M)$ has barycenter x in E. Then there is a compact subset C of M such that $x \in \overline{co}\, C$.

Proof. If p is concentrated on a compact subset C of M, then the assertion reduces to a well-known characterization of the closed convex hull of a compact set (Phelps(1966), prop. 1.2).

If not, due to the tightness of p, there is an increasing sequence of compact sets C_n in M such that

$$C_0 = \emptyset,\ a_n := p(C_{n+1} \setminus C_n) > 0 \quad \text{for each } n \geq 0,$$
$$p(C_n) \uparrow 1 \quad \text{if } n \to \infty.$$

From these, a sufficiently large compact set C can be constructed as follows.

Since $\sum_{n\geq 1} a_n = 1 - a_0 < 1$, we may choose a sequence of numbers b_n such that

$$0 < b_n \leq 1 \quad \text{for every } n \geq 1,$$
$$b_n \to 0 \quad \text{as } n \to \infty,$$
$$\sum_{n\geq 1} b_n^{-1} a_n = 1.$$

Because M is convex, it contains the subset

$$C = C_1 \cup \bigcup_{n\geq 1} (b_n C_{n+1} + (1-b_n)C_1).$$

The set C is in fact compact. Consider a covering $\mathcal{U}$ of C by open subsets of E. Choose a finite subcovering of C_1 and denote its union by U. Again by the compactness of C_1, there is a convex neighbourhood N

of zero such that $U_o := (C_1 + N + N) \subset U$. Because each C_n is contained in the bounded set M and because $b_n \to 0$, there is some $j \geq 1$ such that for every $n \geq j$ we have $b_n C_{n+1} \subset N$ and $(1-b_n)C_1 \subset (C+N)$, hence $b_n C_{n+1} + (1-b_n)C_1$ is contained in the subset U_o of U, which proves that C is compact.
It remains to show that $x \in \overline{co}\, C$. To this end, choose $l \in E'$ and set for each $n \geq 0$

$$c_n := \sup\{l(x) : x \in C_{n+1}\},$$
$$c := \sup\{l(x) : x \in C\}.$$

Then

$$c = \max\left\{c_0, \sup\{b_n c_n + (1-b_n)c_0 : n \geq 1\}\right\} =$$
$$= c_0 + \sup\{b_n(c_n - c_0) : n \geq 1\}.$$

So we have

$$c_n \leq c_0 + b_n^{-1}(c - c_0) \quad \text{for each } n \geq 1,$$
$$l(y) \leq c_n \quad \text{for each } y \in C_{n+1} \setminus C_n.$$

This yields

$$l(x) = \int_M l\, dp = \sum_{n\geq 0} \int_{C_{n+1}\setminus C_n} l\, dp \leq \sum_{n\geq 0} a_n c_n \leq$$
$$\leq a_0 c_0 + \sum_{n\geq 1} a_n (c_0 + b_n^{-1}(c - c_0)) =$$
$$= a_0 c_0 + \left(\sum_{n\geq 1} a_n\right) c_0 + (c - c_0)\left(\sum_{n\geq 1} a_n b_n^{-1}\right) =$$
$$= c_0 + c - c_0 = c.$$

So $l(x) \leq \sup\{l(y) : y \in C\}$ for each $l \in E'$. Since $\overline{co}\, C$ is the intersection of the closed semi-spaces (Schaefer(1971), II.10.1), the proposition is proved. □

<u>1.2.3</u> <u>Corollary</u>. Let M be a bounded subset of a complete locally convex space E. Then for each $p \in P(M)$ the barycenter exists and is

contained in $\overline{\text{co}}$ M.

Proof. Existence of the barycenter was proved in proposition 1.1.3 and the rest follows from the preceding result. □

The following result was originally proved by S.S. Khurana (1969).

1.2.4 *Corollary.* A complete bounded and convex subset of a locally convex space is measure convex.

Proof. Assume that E is embedded in its completion $\tilde{E}$. In $\tilde{E}$ the set in question is bounded convex and closed. The assertion follows now from corollary 1.2.3. □

So the sets for which G. Choquet, G.A. Edgar and others prove integral representation theorems are measure convex (cf. the remarks at the end of this section). Let us consider some more examples of convex sets which are measure convex or not.

Example 2. In example 5 of section 1.1 a completely regular space Y was constructed such that $\mathcal{P}(Y)$ is not measure convex. On the other hand:

(1) $\mathcal{P}(X)$ is measure convex provided that each positive finite τ-smooth measure on X is tight.

Such spaces are called "universally measurable" or "semi-Radonian" - cf. Hoffmann-Jørgensen(1972) and also Sunyach(1979). Locally compact spaces and spaces which can be metrized with a complete metric are semi-Radonian (Topsøe(1970), p. XIII) and also all Radon spaces, where even $P(X) = \mathcal{P}(X)$. So Suslin and in particular Polish spaces have this property.

Proof of (1). For $p \in \mathcal{P}(\mathcal{P}(X))$ define $\mu \in P(X)$ by

$$\mu(B) := \int_{\mathcal{P}(X)} \nu(B)\, dp(\nu) \quad \text{for every } B \in \mathcal{B}(X).$$

If μ is tight then it is the barycenter of p by proposition 1.1.2. By assumption, it is sufficient to show that μ is τ-smooth. To this end, choose a net $(f_i)_{i\in I}$ in $C_b(X)$ filtering downwards and bounded below with limit f. Since every tight measure is τ-smooth we have $\nu(f_i) \downarrow \nu(f)$ for each $\nu \in P(X)$. The functions $\nu \to \nu(f_i)$ in turn are continuous and form a net filtering downwards and bounded below. Since p is also τ-smooth

$$\int f_i \, d\mu = \int \nu(f_i)\, dp(\nu) \downarrow \int \nu(f)\, dp(\nu) = \int f\, d\mu. \qquad \square$$

Example 3. In probability theory there arise in a natural way "moment sets" such as

$$H := \{\nu \in P([0,1]) : \int x d\nu = c_1, \int x^2 d\nu = c_2\}, \; c_i \in \mathbb{R}.$$

More generally: Suppose that X is a semi-Radonian - say Polish - space. For a family $G \subset B(X)$ and real numbers c_g, $g \in G$, define the "generalized moment set"

$$H := \{\nu \in P(X) : \int g d\nu \leq c_g, \; g \in G\}.$$

Using example 2 one reads off from the barycentrical formula in proposition 1.1.2 that H is measure convex.

Let us give a more specific example for a related type of sets. Consider an increasing net of sub-σ-algebras $(F_t)_{t\geq 0}$ of $\mathcal{B}(X)$ and random variables ξ_t. Define

$$H := \{\nu \in P(X) : (\xi_t) \text{ is an } (F_t)\text{-martingale for } \nu\}.$$

H may be rewritten in the form

$$H = \left\{\nu \in P(X) : \{\xi_t : t \geq 0\} \subset L^1(\nu), \int 1_{B_s} (\xi_t - \xi_s)\, d\nu = 0, \; s < t, \; B_s \in F_s\right\}.$$

Under mild regularity conditions H is a measure convex set. The example is from Jacod/Yor(1977).

The importance of such moment sets has been pointed out already by M. Yor (1978). In v. Weizsäcker/Winkler(1979) and (1980) they are studied systematically and integral representation theorems of the Choquet type are proved. The latter reference contains also more examples from diverse parts of probability theory.

Example 4. J.P.R. Christensen and F. Topsøe (1975) asked if a convex G_δ-set in a compact set is measure convex. H. v. Weizsäcker (1976) answered this in the negative, showing that every compact convex metrizable infinite dimensional set contains a convex G_δ-set which is not measure convex.

The following result of D.H. Fremlin and J.D. Pryce ((1974), 2G) characterizes measure convexity as a weak kind of completeness.

1.2.5 Proposition. A subset M of a locally convex space is measure convex if and only if it is bounded and for every compact subset C of M the closed convex hull $\overline{co}$ C is a compact subset of M.

Proof. All of the properties above of M are intrinsic, so we may assume that M is a subset of some complete locally convex space E. Suppose that M is measure convex. Choose a compact subset C of M. By Phelps(1966), proposition 1.2, one has $\overline{co}\ C = r(P(C))$. Hence $\overline{co}$ C is contained in M; because $P(C)$ is compact and the barycenter map r is continuous by proposition 1.1.3, the set $\overline{co}$ C is compact.

Suppose now that M is not bounded. Then there is a neighbourhood U of zero such that for every $i \in \mathbb{N}$ some point x_i from M is not in $2^i U$. The measures $p_n := (\sum_{i=1}^{n} 2^{-i})^{-1} \sum_{i=1}^{n} 2^{-i} \varepsilon_{x_i}$ converge to $p = \sum_{i=1}^{\infty} 2^{-i} \varepsilon_{x_i}$ in $\sigma(M(M), C_b(M))$, hence $r(p_n) \to r(p) = \sum_{i=1}^{\infty} 2^{-i} x_i$, in particular the infinite sum exists. But by construction, $(2^{-i} x_i)_{i \geq 1}$ does not converge to zero, so the series $\sum_{i=1}^{\infty} 2^{-i} x_i$ cannot converge.

Let us now prove the converse. By proposition 1.2.2, each $p \in P(M)$ has its barycenter in $\overline{co}\, C$ for some compact subset C of M and $\overline{co}\, C$ is contained in M by assumption. □

Two stability properties will be needed.

<u>1.2.6</u> <u>Lemma</u>.

a. The intersection of a family of measures convex sets is measure convex.

b. Let $(E_i)_{i \in I}$ be a family of locally convex spaces and denote by E their product space. If for each $i \in I$ the subset M_i of E_i is measure convex then $\prod_{i \in I} M_i \subset E$ is measure convex.

<u>Proof</u>. The first assertion is clear. To prove b., use the equivalence of proposition 1.2.5: choose a compact subset of $\prod_{i \in I} M_i$; each projection C_i of C on the i-th coordinate is a compact subset of M_i, hence also $\overline{co}\, C_i$, and

$$C \subset \prod_{i \in I} \overline{co}\, C_i \subset \prod_{i \in I} M_i .$$

This implies that the closed convex hull of C is a compact subset of $\prod_{i \in I} M_i$. □

Let us conclude this section with a few remarks.

<u>Remarks</u>.

1. By Khurana's result - corollary 1.2.4 - two important cases in integral representation theory are covered by measure convexity:

- classical Choquet theory, which deals with compact convex subsets of locally convex spaces;
- integral representation in closed bounded convex subsets C of Banach spaces, where C has the Radon-Nikodym Property (cf. the appendix).

The second case outlines the setting of Edgar type representation theorems as treated in Edgar(1975), (1976), Bourgin/Edgar(1976), Saint Raymond(1974) and Mankiewicz(1978). The reader is referred to Bourgin (1983). E. Thomas (1980) obtained similar results also for measure convex subsets of locally convex spaces (cf. section 1.5). In v. Weizsäcker(1977) and v. Weizsäcker/Winkler(1980) "conditionally measure convex" sets are considered. The latter property is a slight generalization of measure convexity.

2. Measure convex sets are investigated in Léger/Soury(1971) where they are called "Radons"; in Fremlin/Pryce(1974) they are the bounded Krein sets. These references can be used as an account for measure convexity and related properties; see also Jameson(1972). Edgar(1978) calls measure convex sets "t-convex". We follow Alfsen (1971), p. 130 in the nomenclature.

Section 1.3. Choquet order

One of the main tools in Choquet theory is Choquet order (definition 1.3.3) "as means of determining how 'close' to the 'boundary' of the convex set the various measures under consideration live" (Bourgin(1979)). An exhaustive survey for the case of closed bounded convex sets in Banach spaces is contained in chapter 3 of R.D. Bourgin's book (1983). But this order applies to fairly more general situations - the measures in question may live on unbounded or on non-convex sets. In theorem 1.3.6 below, several characterizations are collected which are useful in this general context and illuminate a variety of its aspects.

Choquet order will be defined on the following set of measures:

<u>1.3.1</u> <u>Definition</u>. Let M be a subset of a locally convex space. Then

$$P'(M) := \{p \in P(M) : l \mid M \in L^1(p) \quad \text{for every } l \in E'\},$$

Remark. If M is bounded, then every linear continuous functional is bounded on M. So $P'(M) = P(M)$ in this case.

The non-specialist may be most familiar with condition a'. in the following proposition.

1.3.2 Proposition. Let M be a convex subset of a locally convex space E. Suppose further that p and q are elements of $P'(M)$. Then the following conditions are equivalent:

a. $\int_M \varphi \, dp \leq \int_M \varphi \, dq$ for every $\varphi \in S(E)$;

a'. $\int_M \varphi \, dp \leq \int_M \varphi \, dq$ for every function $\varphi : M \to \mathbb{R}$ which is convex and continuous;

a". $\int_M \varphi \, dp \leq \int_M \varphi \, dq$ for every function $\varphi : M \to \mathbb{R} \cup \{\infty\}$ which is convex and lower semi-continuous;

(possibly the integrals in a'. and a". may attain the value $+\infty$).

Proof.

Clearly, a" implies a' and a' implies a. So it is sufficient to prove the implication

a. $\Rightarrow$ a". Assume that $\psi : M \to \mathbb{R} \cup \{\infty\}$ is convex and lower semi-continuous. The supergraph

$$G(\psi) := \{(x,a) \in M \times \mathbb{R} : \psi(x) \leq a\}$$

is convex and its closure $G(\psi)^a$ in $E \times \mathbb{R}$ is closed and convex. Choose $x \in M$ and $a \in \mathbb{R}$ such that $\psi(x) > a$. Then $(x,a) \notin G(\psi)^a$. By the geometrical version of the Hahn-Banach theorem there is $l \in E'$ and $c \in \mathbb{R}$ such that

$$l(x) + ca < \inf\{l(x') + ca' : (x',a') \in G(\psi)^a\}.$$

Necessarily $c > 0$, since for $x' \in M$ the set $\{a' \in \mathbb{R} : (x',a') \in G(\psi)^a\}$ is unbounded above whenever $\psi(x') < \infty$; but we may assume that there is some $x' \in M$ with $\psi(x') < \infty$.

Set $k := -c^{-1}l$ and $b := c^{-1}l(x) + a$; then

$$k(x) + b = a$$

and for every $x' \in M$

$$\begin{aligned} k(x') + b &= -c^{-1}l(x') + c^{-1}(l(x) + ca) < \\ &< -c^{-1}l(x') + c^{-1}\inf\{l(x') + ca' : (x',a') \in G(\psi)^a\} \leq \\ &\leq \psi(x'). \end{aligned}$$

Hence ψ is the pointwise sumpremum of all $\varphi \in S(E)$ which are below ψ on M. The latter form an increasing net, p and q are τ-smooth (cf. 0.4), so a. implies

$$\int_M \psi \, dp \leq \int_M \psi \, dq. \qquad \square$$

Note that condition a. may be formulated even when M is not convex. We use it to define Choquet order.

1.3.3 Definition. Let M be a subset of a locally convex space E. The relation $<$ on $P'(M)$ defined by

$$p < q \text{ iff } \int \varphi \, dp \leq \int \varphi \, dq \quad \text{for every } \varphi \in S(M)$$

is called Choquet order.

In fact, the Choquet order is an ordering.

1.3.4 Lemma. The relation $<$ from definition 1.3.3 is reflexive, antisymmetric and transitive.

Proof. Antisymmetry follows from part b. of the lemma in section 0.7.
□

The name "Choquet order" is somehow misleading. This order goes back to Hardy, Littlewood and Polya and was used by Bohnenblust, Shapley, Sherman, Blackwell and many others for the comparison of sampling procedures (cf. Blackwell(1951) and respective references in Heyer (1982)). It is also well-known as "balayage order", cf. Doob(1968). G. Choquet initiated its use in integral representation theory.

Two basic concepts from probability theory will be needed in theorem 1.3.6.

<u>1.3.5</u> <u>Definition</u>. Let M be a subset of a locally convex space E.

a. Suppose that $p \in P(M)$. A mapping $T : M \to P(M)$, $x \to T_x$ is called a <u>p-dilation</u> iff

(i) the mapping $x \to T_x(B)$ is p-measurable for each $B \in \mathcal{B}(M)$,

(ii) for each $l \in E'$, we have for p-almost all $x \in M$ that

$$l \mid M \text{ is p-integrable,}$$

$$l(x) = \int_M l \, dT_x .$$

The measure $B \to \int_M T_x(B) \, dp(x)$ on $\mathcal{B}(M)$ is denoted by pT.

b. Let $(X, F, \mathbb{P})$ be a probability space and G a sub-σ-algebra of F. Suppose further that $f : X \to M$ is an F-$\mathcal{B}(M)$ and $g : X \to M$ is a G-$\mathcal{B}(M)$ measurable mapping. We call g <u>conditional expectation</u> of f given G and write $g = \mathbb{E}(f \mid G)$ iff

$$l \circ f \in L^1(\mathbb{P}) \text{ and } l \circ g \in L^1(\mathbb{P} \mid G) \quad \text{for every } l \in E',$$

and

$$\int_G l \circ f \, d\mathbb{P} = \int_G l \circ g \, d\mathbb{P} \quad \text{for every } G \in G \text{ and } l \in E'.$$

<u>Remarks</u>.

1. A cumbersome feature of p-dilations is the following: condition (ii) does *not* say, that each x is the barycenter of T_x, since (ii) may fail for each $l \in E'$ on an individual null-set. If M is a Suslin set, the situation becomes considerably more conspicuous.

Then, to any p-dilation T there is an associated <u>dilation</u> T' - i.e.

$x \to T'_x(B)$ is Borel measurable for each $B \in \mathcal{B}(M)$ and every $x \in M$ is the barycenter of T'_x - and a p-null-set $N \in \mathcal{B}(M)$ such that $T_x = T'_x$ for every $x \in M \setminus N$.

The proof for this fact is - up to a minor modification - the same as for proposition 1.4.2.c. below.

This allows the following interpretation: For p-almost all x the p-dilation T sweeps out the point mass in x to the distribution T_x with barycenter x.

2. Definition b. says that for every $l \in E'$ the function $l \circ g$ is the usual (one-dimensional) conditional expectation of $l \circ f$ given G, which exists as soon as $l \circ f$ is integrable and which is unique up to a set of measure zero.

Let us now enlist some equivalent forms of Choquet order.

1.3.6 Theorem. Let M be a subset of the locally convex space E. Suppose that p and q are elements of $P'(M)$. Then the following conditions a.-d. are equivalent:

a. $\int \varphi \, dp \leq \int \varphi \, dq$ for each $\varphi \in S(M)$;

b. there is $\mu \in P(M \times M)$ with marginal measures p on the first and q on the second coordinate such that for every $B \in \mathcal{B}(M)$ and $l \in E'$

$$\int_{B \times M} l \circ pr_1 \, d\mu = \int_{B \times M} l \circ pr_2 \, d\mu ,$$

where pr_i denotes the i-th projection;

c. there are a probability space $(X, F, \mathbb{P})$, a sub-σ-algebra G of F and measurable mappings g and f from X into M such that

$$p = \mathbb{P} \circ g^{-1}, \; q = \mathbb{P} \circ f^{-1} \quad \text{and} \quad g = \mathbb{E}(f \mid G);$$

d. there is a p-dilation T such that $pT = q$.

Assume moreover that M is bounded. Then condition a.-d. are also equivalent to

e. there is $\mu \in P(M \times M)$ such that

(i) $q(B) = \int_{P(M)} m(B)\, d\mu(m)$ for every $B \in \mathcal{B}(M)$,

(ii) $p(B \cap M) = \mu\{m \in P(M) : r(m) \in B\}$ for every $B \in \mathcal{B}(E)$.

We will prove this theorem after some remarks.

Remarks.

1. Condition d. is the analogue of the dilation-order, well-known in the compact case. Condition c. (which is essentially the same as b.) was introduced by G.A. Edgar (1976). But it had already been used by J.L. Doob (1968) and also appears in Hoffmann-Jørgensen(1977), p. 44. In the setting of closed bounded convex subsets of Banach spaces Edgar(1976) used the dilation order in connection with an existence theorem for integral representation, Bourgin/Edgar(1976) used c. and d. to study uniqueness and Saint Raymond(1974) used a. also for uniqueness problems. The just mentioned orderings are reviewed in Bourgin(1979) and later in Bourgin(1983).

2. Condition e. is due to H. v. Weizsäcker (1977). It may be considered as a converse to condition d. As the latter says how to get - given p - all q greater than p in Choquet order (smear out point masses and mix by means of p), condition e. gives a recipe how to construct - given q - all p smaller in Choquet order (decompose q in all possible ways to get μ's and take the image measure by means of the barycenter map). It is open, if the assumption that H is bounded may be dropped.

3. Parts of theorem 1.3.6 are proved in several papers:
a.⇔b.⇔c.⇔d. in v. Weizsäcker / Winkler(1980); for bounded sets
a.⇔b.⇔c.⇔d.⇔e. in v. Weizsäcker(1977) and for bounded convex sets
a.⇔c.⇔d. in Edgar(1978).

Proof of theorem 1.3.6. We will prove the implications a.⇒b.⇔c. and b.⇒d.⇒a. in the general case and for bounded M the implications d.⇒e.⇒a.

"a.⇒b.": Consider the linear space

$$V := \{f \in C(H \times H) \; : \; |f(x,y)| \leq g(x) + h(y) \text{ for some continuous } g,\, h \in L^1(p+q)\}.$$

Whenever g and h are functions on M let $g \otimes h(x,y) := g(x)h(y)$. Define a linear subspace of V:

$$W := \{g \otimes 1 + 1 \otimes h \; : \; g,\, h \in L^1(p+q) \cap C(M)\}$$

and a linear functional k on W by means of p and q:

$$k(g \otimes 1 + 1 \otimes h) := \int g\, dp + \int h\, dq.$$

The aim is to extend k to all of V and then to show that the extension induces the desired measure μ.

To see that k may be extended, we verify the assumptions of a generalized version of the Hahn-Banach theorem from Dinges(1970), E.5.4 on page 468; it reads as follows:

Let $(V,\leq)$ be an ordered linear space, generated by its positive elements. Let ρ be a sublinear functional on V, with $-\infty < \rho(f) \leq \infty$ for $f \in V$ and such that $\rho(f) \leq 0$ whenever $f \leq 0$. Suppose further that W is a linear subspace of V and that for each $f \in V$ there are elements g, $h \in W$ with $g \leq f \leq h$. Then any linear functional k on W which is below ρ can be extended to a positive linear functional m on V below ρ.

With the natural order, V and W defined above fulfill the hypothesis.

Next ρ has to be found. For $f \in V$ set

$$\hat{f}(x) := \inf\{\varphi(x) \; : \; -\varphi \in S(E),\; \varphi(y) \geq f(x,y) \quad \text{for every } y \in M\}$$

to get $\hat{f} : M \to \mathbb{R} \cup \{\infty\}$ and

$$\rho : V \to \mathbb{R} \cup \{\infty\},\ f \to \int^* \hat{f}(x)\,dp(x),$$

where $\int^* \cdot\, dp$ is the upper integral associated with p.

The functional ρ is sublinear and $f \leq 0$ implies $\rho(f) \leq 0$; furthermore k is below ρ on W:

$$\begin{aligned} k(g \otimes 1 + 1 \otimes h) &= \int\int \Big(g(x) + h(y)\Big) dp(x)\,dq(y) \leq \\ &\leq \int\int^* \widehat{g\otimes 1 + 1\otimes h(x)}\, dp(x)\,dq(y) = \\ &= \rho(g \otimes 1 + 1 \otimes h). \end{aligned}$$

The above Hahn-Banach theorem yields a positive linear form $m : V \to \mathbb{R}$ below ρ with the properties

(1) $\qquad m(g \otimes 1) = \int g\, dp,$

(2) $\qquad m(1 \otimes h) = \int h\, dq.$

Furthermore, for each bounded continuous function g on M and each $\varphi \in -S(E)$, the function $g \cdot \varphi$ is continuous and in $L^1(p+q)$ and

$$\begin{aligned} m(g \otimes \varphi) \leq \rho(g \otimes \varphi) &= \int g(x)\varphi(x)\, dp(x) = \\ &= m(g\varphi \otimes 1). \end{aligned}$$

Since $E' \subset (S(E) \cap -S(E))$

(3) $\qquad m(gl \otimes 1) = m(g \otimes l).$

The equalities (1) - (3) would imply b. if there were $\mu \in P(H \times H)$ such that $m(f) = \int f\, d\mu$ for each $f \in V$. But for that, it is sufficient that the marginals p and q of m are tight (use e.g. Fremlin/Garling/Haydon (1972), thm. 1). Thus b. holds.

Remark. The Hahn-Banach version of H. Dinges (1970) was used by H. Rost (1971) to establish a similar relation between conical measures.

Proof of theorem 1.3.6 continued.

"b.$\Leftrightarrow$c.": Condition b. may be rephrased as follows: for $\mu \in P(M \times M)$ the first projection of $M \times M$ is the conditional expectation of the second, given the σ-algebra generated by the first, and the image measures of μ under the projections are p and q respectively. Hence c. holds. Conversely, assume that c. is valid. The mapping $x \to (g(x), f(x))$ is measurable w.r.t. the product σ-algebra $\mathcal{B}(M) \otimes \mathcal{B}(M)$ (consider first measurable rectangles). Set for each $B \in \mathcal{B}(M) \otimes \mathcal{B}(M)$

$$\mu'(B) := \mathbb{P}\{x \in X : (g(x), f(x) \in B\}.$$

The measure μ' has the prescribed marginal measures p and q. Extend now μ' to $\mu \in P(M \times M)$ to get b.

"b.⇒d.": This is an easy consequence of v. Weizsäcker's disintegration theorem A2 in the appendix.

Assume that b. holds. Because p is the first marginal measure of μ, we have $\mathcal{B}_p(M) \otimes \mathcal{B}(M) \subset \mathcal{B}_\mu(M \times M)$; μ restricted to $\mathcal{B}_p(M) \otimes \mathcal{B}(M)$ has the properties of $\mathbb{P}$ in A2. Apply this result to get a mapping

$$T : M \to P(M), \; x \to T_x$$

which is denoted by v there.

By hypothesis and property A2. (iii), we have for each $l \in E'$ and $B \in \mathcal{B}(M)$ that

$$\begin{aligned} &\int_B l(x)\, dp(x) = \\ &= \int_{B \times M} l(x)\, d\mu(x,y) = \\ &= \int_{B \times M} l(y)\, d\mu(x,y) = \\ &= \int_B \left[\int_M l\,(y)\, dT_x(y) \right] dp(x). \end{aligned}$$

Hence for each $l \in E'$

$$l(x) = \int_M l(y)\, dT_x(y) \quad \text{for p-almost all } x \in M.$$

This shows that T is a p-dilation and finally the disintegration theorem A2 assures that $pT = \mu \circ pr_2^{-1} = q$ and $p \circ T^{-1}$ is tight. Thus d. is proved.

"d.⇒a.": If T is a p-dilation such that $q = pT$, we get - neglecting p-zero-sets - for $l_i \in E'$ and $c_i \in \mathbb{R}$ that

$$\int_M \max_{1 \leq i \leq n} (l_i(x) + c_i)\, dp(x)$$

$$= \int_M \max_{1 \leq i \leq n} \left[\int_M l_i(y)\, dT_x(y) + c_i\right] dp(x) \leq$$

$$\leq \int_M \left[\int_M \max_{1 \leq i \leq n} (l_i(y) + c_i)\, dT_x(y)\right] dp(x) =$$

$$= \int_M \max_{1 \leq i \leq n} (l_i(x) + c_i)\, dq(x)$$

which proves a.

Assume now in addition that M is bounded.

"d.⇒e.": By assumption, there is a p-dilation T with $q = pT$ and tight image measure $\mu := p \circ T^{-1}$ on $P(M)$.

From the definitions, we have

$$q(B) = \int_M T_x(B)\, dp(x) = \int_{P(M)} m(B)\, d\mu(m) \quad \text{for every } B \in \mathcal{B}(M),$$

hence e.(i) is true.

Since M is bounded, by proposition 1.1.3 the barycenter map r is defined on all of $P(M)$ with values in the completion $\tilde{E}$ of E, and since it is continuous also by 1.1.3, it is measurable. Let us think of E as a subset of $\tilde{E}$. Choose $B \in Z(\tilde{E})$; there is a countable subset G of $\tilde{E}'$ and $B^* \in \mathcal{B}(\mathbb{R}^G)$ such that

$$B = \{x \in \tilde{E} : (l(x))_{l \in G} \in B^*\},$$

hence

$$\mu(\{m \in P(M) : r(m) \in B\}) =$$

$$= p(\{x \in M : r(T_x) \in B\}) =$$

$$= p(\{x \in M : \left(l(r(T_x))\right)_{l \in G} \in B^*\}) =$$

$$= p(\{x \in M : (l(x))_{l \in G} \in B^*\}) =$$

$$= p(M \cap B).$$

In consequence, the two measures from e.(ii) act equally on $S(M)$, since $\tilde{E}'$ may be identified with E'. Hence they coincide because of the lemma in 0.7. This proves e.(ii).

"e.⇒a.": If on the other hand e. holds, we have for

$\varphi = \max_{1 \le i \le n} (l_i(\cdot) + c_i) \in S(M)$ that

$$\int \varphi \, dp =$$

$$= \int_{P(M)} \max_{1 \le i \le n} [l_i(r(m)) + c_i] \, d\mu(m) =$$

$$= \int_{P(M)} \max_{1 \le i \le n} \left[\int_M l_i(y) \, dm(y) + c_i\right] d\mu(m) \le$$

$$\le \int_{P(M)} \left[\int_M \max_{1 \le i \le n} (l_i(y) + c_i) \, dm(y)\right] d\mu(m) =$$

$$= \int_M \varphi \, dq.$$

This is condition a.

Now the proof of theorem 1.3.6 is complete. □

Section 1.4. Boundary measures

A genuine integral representation theorem of Choquet type has the form

> If the convex set M enjoys property ( ) then each element $x \in M$ is the barycenter of some $p \in P(\text{ex } M)$.

It is well known that even when M is compact, difficulties arise as soon as M is not separable (cf. the remarks at the end of this section). A natural and appropriate substitute for tight measures *on* the extreme points are measures which are maximal in Choquet order. Intuitively, their mass is swept outside as far as possible.

<u>1.4.1</u> <u>Definition</u>. Let M be a subset of a locally convex space. A measure $p \in P'(M)$ is called a <u>boundary measure</u> on M iff it is maximal in Choquet order.

In many special cases the name "boundary measure" can be justified to a certain extent. But it may also fail in the worst sense to be literally true. G.A. Edgar (1977) (unpublished) gave an example of a closed bounded convex - hence measure convex - subset of a Banach space which admits a boundary measure but has no extreme point. The same example was independently found by P. Mankiewicz (Edgar(1983), personal communication). M. Talagrand (1982) improved Edgar's result considerably, constructing a closed bounded convex subset of $c_o(\mathbb{R})$ which has no extreme point but every element is the barycenter of a boundary measure. It is contained in Bourgin(1983), 6.5.5.

Positive results are:

<u>1.4.2</u> <u>Proposition</u>. Let C be a convex subset of a locally convex space E. Then:

a. Assume that either

$r : P(C) \to C$ is open and C is compact

or

$r : P(C) \to C$ is open and C is bounded and completely metrizable.

Then for every boundary measure p on C we have $p^*(\text{ex } C) = 1$.

b. If C is bounded and Suslin, then ex C is universally measurable and $p(\text{ex } C) = 1$ for every boundary measure p on C.

c. Assume that C is measure convex and Suslin. Then $p \in P(C)$ is a boundary measure if and only if $p(\text{ex } C) = 1$.

Proof.

a. Since this result will not be used in the sequel, the reader is referred to Edgar(1978), proposition 2.5, (ii) and (iii).

All arguments used in the proofs of b. and c. are hidden in v. Weizsäcker/Winkler(1980).

b. A Suslin space is strongly Lindelöf (Schwartz(1973), prop. 3 on p. 104); we can apply proposition 4 of Schwartz(1973), p. 105, to get a sequence $(l_i)_{i\in\mathbb{N}}$ in E' which separates points in C. Furthermore, we can choose suitable positive real numbers a_i such that

$$\psi(x) := \sum_{i\in\mathbb{N}} a_i (l_i(x))^2$$

defines a bounded strictly convex function on C.

Since C is Suslin, $C\times C$ is Suslin and also its open subset $G := \{(x,y)\in C\times C : x \neq y\}$ (Schwartz(1973), thm. 3 on p. 96).
The mapping

$$h : G \to C,\ (x,y) \to \frac{1}{2}(x+y)$$

is continuous; thus its image $C\setminus \mathrm{ex}\, C$ is Suslin. By corollary III.11.7 in Hoffmann-Jørgensen(1979) there is a right inverse $f : C\setminus \mathrm{ex}\, C \to G$ of h measurable w.r.t. the universally measurable sets. Extend f to C, setting $f(x) := (x,x)$ if $x\in \mathrm{ex}\, C$ to get a universally measurable function from C to $C\times C$. The component functions f_1 and f_2 are universally measurable and

$$\frac{1}{2}(f_1(x) + f_2(x)) = x \quad \text{for every } x\in C,$$
$$f_1(x) \neq f_2(x) \quad \text{whenever } x\in C\setminus \mathrm{ex}\, C.$$

Let now p be a boundary measure on C and assume that $p(\mathrm{ex}\, C) < 1$. Then there is a compact set $K\subset (C\setminus \mathrm{ex}\, C)$ such that f_1 and f_2 are continuous on K and $p(K) > 0$ because of Lusin's theorem. Define

$$T_x := \varepsilon_x \text{ for } x\notin K,\quad T_x := \frac{1}{2}(\varepsilon_{f_1(x)} + \varepsilon_{f_2(x)}) \quad \text{for } x\in K.$$

Then

$$\int_C \psi \, dpT - \int_C \psi \, dp = \int_K \left(\frac{1}{2}[\psi(f_1(x)) + \psi(f_2(x))] - \psi(x) \right) dp > 0 .$$

Hence $p \neq pT$, but by theorem 1.3.6.d. we have $p < pT$, contradicting the assumption that p is maximal in Choquet order.

c. Assume now that C is Suslin and measure convex, p(ex C) = 1 and $p < q$. According to theorem 1.3.6.d. there is a p-dilation T such that $q = pT$.

$P(C)$ is a Suslin space (Bourbaki(1969), 5.4.2) - hence also its image K under the continuous mapping r. There is a countable subset G of E' separating points of K (Schwartz(1973), prop. 4 on p. 105). Since T is a p-dilation, we have

$$l(x) = \int l(y) \, dT_x(y) = l(r(T_x))$$

p-almost everywhere for all $l \in G$, i.e. $x = r(T_x)$ p-almost everywhere. An extreme point of a measure convex set can only be represented by its point measure, which will be shown in proposition 1.5.4. Moreover p is concentrated on the extreme points of C, hence $T_x = \varepsilon_x$ for p-almost all x. This implies $p = pT = q$. □

Remark. The second part of proposition 1.4.2.b. is contained in Edgar(1978) as part (i) of proposition 2.5. In the proof Edgar falsely claims that each bounded convex Suslin subset of a locally convex space is measure convex. As example 1 in section 1.2. shows, this is not true.

Two examples may justify the hypothesis' in proposition 1.4.2.

Example 1. By the Choquet-Bishop-de Leeuw theorem for convex compact sets each boundary measure vanishes on Baire subsets of C disjoint from ex C (Phelps(1966), p. 27). The same authors provided the classical examples of a compact convex set C, such that ex C is a Borel set and a boundary measure p on C with p(ex C) = 0 (Bishop/de Leeuw(1959)).

Example 2. We construct a σ-compact metrizable convex subset H of $P([0,1]^2)$ with an extreme point x such that $\varepsilon_x < p$ and $\varepsilon_x < q$ for two different boundary measures p and q. Furthermore, every point in H has at least one representing boundary measure. In particular ε_x is no boundary measure despite of $\varepsilon_x(\text{ex } H) = 1$ and H is Suslin as the countable union of Suslin spaces; clearly H cannot be measure convex.

Let H denote the convex hull of the set

$$C := \{\varepsilon_s \otimes \lambda : s \in [0,1]\} \cup \{\lambda \otimes \varepsilon_t : t \in [0,1]\} \cup \{\lambda^2\}$$

in $P([0,1]^2)$, where λ denotes Lebesgue measure on [0,1]. C is a closed subset of $P([0,1]^2)$, which is compact and metrizable. So the sets

$$H_n := \{\sum_{i=1}^{n} a_i v_i : a_i \geq 0, \sum_{i=1}^{n} a_i = 1, v_1,\ldots,v_n \in C\}$$

are compact and metrizable. Since the representations $\sum_{i=1}^{n} a_i v_i$ are unique, we have C = ex H. The extreme point λ^2 has the two different barycentric representations

$$\begin{aligned}&\lambda^2(B) = \\ &= \int_0^1 \varepsilon_s \otimes \lambda\,(B)\, d\lambda(s) = \\ &= \int_0^1 \lambda \otimes \varepsilon_t\,(B)\, d\lambda(t) \quad \text{for each } B \in \mathcal{B}([0,1]^2).\end{aligned}$$

Define maps

$$\theta_1 : [0,1] \to H,\ s \to \varepsilon_s \otimes \lambda,$$
$$\theta_2 : [0,1] \to H,\ t \to \lambda \otimes \varepsilon_t.$$

The measure $p := \lambda \circ \theta_1^{-1}$ is concentrated on $\{\varepsilon_s \otimes \lambda : s \in [0,1]\}$, the measure $q := \lambda \circ \theta_2^{-1}$ is concentrated on $\{\lambda \otimes \varepsilon_t : t \in [0,1]\}$.

Both measures have barycenter λ^2. The set

$$H^* := \{\alpha(\mu \otimes \lambda) + (1-\alpha)\ (\lambda \otimes \nu)\ :\ 0 \leq \alpha \leq 1,\ \mu, \nu \in P([0,1])\}$$

is convex compact and metrizable, further

$$\left(\{\varepsilon_s \otimes \lambda\ :\ s \in [0,1]\} \cup \{\lambda \otimes \varepsilon_t\ :\ t \in [0,1]\}\right) \subset \text{ex } H^*.$$

By proposition 1.4.c, the measures p and q considered as elements of $P(H^*)$ are maximal in Choquet order and all the more they are boundary measures on H.

We note two geometrical properties of the set of boundary measures.

<u>1.4.3</u> <u>Proposition</u>. Let M be a subset of a locally convex space E. The set of boundary measures on M is convex.

<u>Proof</u>. Choose boundary measures p_1 and p_2 on M. Suppose $p := \frac{1}{2}(p_1+p_2) \prec q$ for $q \in P'(M)$. By theorem 1.3.6 there is a p-dilation T with $pT = q$. The measures p_1T and p_2T are also in $P'(M)$, since

$$\int l\, dp_i \leq 2\int l\, dq \quad \text{for } i = 1,2 \quad \text{and} \quad l \in E'.$$

Moreover $p_1 \prec p_1T$ and $p_2 \prec p_2T$; since p_1 and p_2 are maximal in Choquet order this implies $p_1T = p_1$ and $p_2T = p_2$. Consequently $q = \frac{1}{2}(p_1T+p_2T) = p$, hence p is a boundary measure. □

Recall that a subset C of a linear space H is called a <u>pointed convex cone</u> iff

(i) C is a convex cone;

(ii) if x and -x are in C then $x = 0$;

(iii) C contains the origin and a nonzero vector.

Such a cone C induces a (reflexive, antisymmetric and transitive) order $\leq_C$ in H defined by

$$x \leq_C y \quad \text{iff} \quad y - x \in C \quad \text{for } x, y \in H.$$

C is called a <u>lattice cone in its own order</u> iff $C - C$ is a vector lattice w.r.t. $\leq_C$.

<u>1.4.4</u> <u>Proposition</u>. Let M be a subset of a locally convex space. The cone $\{\alpha p : \alpha \geq 0,$ p boundary measure on M$\}$ generated by the boundary measures is a lattice cone in its own order.

<u>Proof</u>. Denote the set of boundary measures on M by $P^b(M)$. Since $M_+(M)$ is a lattice cone in its own order, we have only to verify that $p \in P(M)$, $q \in P^b(M)$ and $p \leq \alpha q$ for some $\alpha > 0$ implies $p \in P^b(M)$ (then the minimum of two elements of $\mathbb{R}_+ \cdot P^b(M)$ taken in $M_+(M)$ is already the minimum in $\mathbb{R}_+ \cdot P(M)$). Assume that $p \prec p'$ for some $p' \in P(M)$. The measure $q' := q - \alpha^{-1}p + \alpha^{-1}p'$ is in $P(M)$ and dominates q in Choquet order, hence $q = q'$ by assumption and finally $p = p'$. So p is a boundary measure.
□

We conclude this section with some

<u>Remarks on integral representation by means of boundary measures</u>.

1. The celebrated Choquet theorem reads:

Let C be a compact convex metrizable subset of a locally convex space. Then ex C is a G_δ-set and for every $x \in C$ there is an element $p \in P(\text{ex } C)$ which represents x.

If metrizability is dropped, the extreme boundary is in general no Borel set. Moreover, there may be an element x which has no representing measure p such that $p^*(\text{ex } C) = 1$. This is shown by examples in Bishop/ de Leeuw(1959).
Hence one is looking for a representation by means of measures whose mass is as close near the extreme boundary as possible. These are - in view of the interpretation of Choquet order - the boundary measures. Choquet's theorem now becomes:

Let C be a compact convex subset of a locally convex space. Then for every element x of C there is a *boundary* measure which represents x.

If in general there is no representing tight probability measure on the extreme points, one may ask if there is at least a representing probability measure m on some σ-algebra $\mathcal{B}'$ on C, weaker then $\mathcal{B}(C)$, with $m^*(\text{ex } C) = 1$. The Choquet-Bishop-de Leeuw theorem shows that Baire sets are appropriate (Phelps(1966)):

Let C be a compact subset of a locally convex space. Then for every element x of C there is a probability measure p on the Baire-σ-algebra of C, fulfilling the barycentrical formula and $p^*(\text{ex } C) = 1$.

One can follow these lines to get similar results in other cases.

Let us only mention an example from measure theory:

Let C be a closed convex bounded subset of a cone $M_+(X)$ or a moment set (cf. example 3 in section 1.2). Endow C with the evaluation σ-algebra $\Sigma(C)$ generated by the maps $C \ni \nu \to \nu(B)$, $B \in \mathcal{B}(X)$. Then each $D \in \Sigma(C)$ disjoint from the extreme boundary has measure zero w.r.t. each boundary measure on C.

So a representation theorem of the Choquet-Bishop-de Leeuw type is obtained (Winkler(1980)). Unfortunately such approaches do not work, if one is interested in *uniqueness* of representing measures. Then separability assumptions seem to be inevitable (cf. the remarks following 1.5.6).

2. Already G. Choquet started two approaches to get rid of the compactness assumption: "conical measures" and "caps". A milestone is Edgar's representation theorem for subsets of Banach spaces which are closed bounded convex and have the Radon-Nikody Property (Edgar(1975)) - see also the concluding remark in section 1.2. Partial success was

already achieved by Bourgin(1971). Results of that type are of considerable interest because of their connection with the geometry of Banach spaces. These aspects are described in detail in Bourgin(1983). Later on, a variety of authors obtained results along these lines - see the references in section 1.2, in Fuchssteiner/Lusky(1981) and in Bourgin(1983) - among them the important work of E. Thomas. Noncompact integral representation theorems are also contained in a series of papers by v. Weizsäcker and the author between 1975 and 1980 listed in the references of this paper.

Section 1.5. Simplices

There are several current definitions of a simplex, which in general do not coincide. We will study stability properties of sets which allow unique integral representations by means of boundary measures. So the following definition is the natural one in the present context:

1.5.1 Definition. Let S be a measure convex subset of a locally convex space. S is called a simplex iff every element x of S is the barycenter of one and only one boundary measure p_x on S.

Near at hand is

Example 1. Let X denote a completely regular Suslin space. The space $P(X)$ of tight probability measures is measure convex by example 2 in section 1.2. The barycenter map is a bijection between $P(X)$ and $P(\mathrm{ex}\, P(X)) = P(\{\varepsilon_x : x \in X\})$ according to example 3 in section 1.1. Since $P(X)$ is a Suslin space (Bourbaki(1969), cor. 5.4.2 on p. 62), proposition 1.4.2.c. tells us that $P(X)$ is a simplex.

According to the Choquet-Meyer theorem, a compact convex set is a simplex (in our sense) if and only if it fulfils the lattice theoretic condition formulated in definition 1.5.2 below. G. Choquet used the latter to *define* a simplex (cf. the remarks and historical note below).

1.5.2 Definition. A convex subset K of a linear space H is called a Choquet simplex iff it enjoys the property

(L) the cone $\mathbb{R}_+ \cdot (K \times \{1\}) \subset (H \times \mathbb{R})$ is a lattice cone in its own order.

Let us compare these two notions of a simplex.

1.5.3 Proposition. Each simplex is a measure convex Choquet simplex.

Proof. Denote the simplex in question by S and the set of boundary measures by $P^b(S)$. By proposition 1.4.4 the boundary measures generate a lattice cone in its own order. The map

$$P^b(S) \ni p \to (r(p),1) \in S \times \{1\}$$

is an affine bijection, hence $\mathbb{R}_+ \cdot (S \times \{1\})$ is also a lattice cone in its own order. So S is a Choquet simplex. □

The converse does not hold. It can break down because of several reasons.

Example 2. Denote the Lebesgue measure on [0,1] by λ. The set

$$TC := \{\nu \in P([0,1]) : \nu \text{ is totally continuous w.r.t. } \lambda\}$$

is clearly a Choquet simplex (the own order of the generated cone is the setwise order on measures). TC is also measure convex.
On the other hand, TC is a Lusin space in the weak topology; in fact, there is a stronger Polish topology on TC induced by the map

$$\Theta : TC \to L^1([0,1],\lambda), \ \nu \to \frac{d\nu}{d\lambda}$$

where $\frac{d\nu}{d\lambda}$ denotes the Radon-Nikodym derivative. It is well-known, that TC has no extreme points. So proposition 1.4.2.b. implies that there is no boundary measure on TC, hence it is no simplex.

Example 3. The set D := co $\{\varepsilon_x : x \subset [0,1]\}$ of discrete probability measures is a Choquet simplex and each element has a unique representing boundary measure. D is not measure convex.

Example 4. In example 1, the Choquet simplex TC was no simplex, because there is a lack of boundary measures. The opposite may also happen. Recall the set H from example 2 in section 1.4. Every element of H is the barycenter of a unique discrete probability measure on the extreme points. Like in example 3, the latter form a Choquet simplex, hence H is one too. But for the extreme point λ^2 there are two different representing boundary measures.
Note that ε_{λ^2} is concentrated on the extreme points of H but no boundary measure. Since H is even a Lusin set, H cannot be measure convex according to proposition 1.4.2. We will show in corollary 1.5.5, that such pathologies cannot occur if the set in question is measure convex.

The following technical statement from v. Weizsäcker/Winkler(1980) is needed.

1.5.4 Lemma. Let M denote a measure convex subset of some locally convex space E. Assume that $x \in M$, $p \in P(M)$ and that $x = r(p)$. Then for every $\varphi \in S(M)$ there is a discrete probability measure $m = \sum_{i=1}^{n} a_i \varepsilon_{x_i}$ on M and there are elements $p_i \in P(M)$ such that

(i) $x = r(m)$, $x_i = r(p_i)$, $p = \sum_{i=1}^{n} a_i p_i$,

(ii) $\int \varphi \, dp = \int \varphi \, dm$.

Proof. Recall that φ has the form $\varphi(x) = \max\{l_i(x) + c_i : 1 \le i \le r\}$ with $l_i \in E'$ and $c_i \in \mathbb{R}$. There are disjoint Borel sets $B_1, \ldots, B_r$ in M such that $\varphi | B_i = (l_i + c_i) | B_i$. Assume that $a_i := p(B_i) > 0$ exactly for the first n of these sets, so that $\sum_{i=1}^{n} a_i = 1$. Set now

$$p_i := \frac{1}{a_i}\, p(\cdot \cap B_i), \quad 1 \leq i \leq n.$$

Then $p_i \in P(M)$, $p = \sum_{i=1}^{n} a_i p_i$ and $x_i := r(p_i) \in M$ since M is measure convex. Define

$$m := \sum_{i=1}^{n} a_i \varepsilon_{x_i};$$

then

$$x = r(p) = r\Big(\sum_{i=1}^{n} a_i p_i\Big) = \sum_{i=1}^{n} a_i r(p_i) = \sum_{i=1}^{n} a_i x_i = r(m).$$

Moreover

$$\int \varphi\, dp = \sum_{i=1}^{n} a_i \int \varphi\, dp_i = \sum_{i=1}^{n} a_i \varphi(x_i) = \int \varphi\, dm. \qquad \square$$

The following result was announced in example 4.

<u>1.5.5 Corollary</u>. Suppose that M is a measure convex subset of some locally convex space. Then each extreme point x of M has exactly one representing measure, namely the point measure ε_x. In particular, ε_x is a boundary measure.

<u>Proof</u>. Assume $x \in \text{ex}\, M$ and $x = r(p)$ for $p \in P(M)$. In lemma 1.5.4, we must have $n = 1$, so $p(\varphi) = \varphi(x)$ for every $\varphi \in S(M)$. Consequently $p = \varepsilon_x$. The rest is clear. $\square$

Since in a simplex every element has a unique representing boundary measure, each point measure ε_x is dominated in Choquet order by one and only one boundary measure p_x on S. More generally

<u>1.5.6 Proposition</u>. On a simplex S, every element of $P(S)$ is majorized in Choquet order by one and only one boundary measure.

<u>Proof</u>. Let for every $y \in S$ the unique representing boundary measure be denoted by p_y. Choose $p \in P(S)$ and set $x := r(p)$. Let further $m = \sum_{i \leq n} a_i \varepsilon_{x_i}$ be a discrete probability measure with barycenter x. The

measure $m' := \sum_{i \leq n} a_i p_{x_i}$ is a boundary measure, because the set $P^b(S)$ of boundary measures is convex by proposition 1.4.3. Clearly m' dominates m. Since $r(m') = x = r(p)$, we have $m' = p_x$, hence p_x dominates m.

Choose now $\varphi \in S(S)$. Lemma 1.5.4 provides a discrete probability measure m with barycenter x and $p(\varphi) = m(\varphi)$. We have shown above that p_x dominates every discrete probability measure with barycenter x, so

$$\int \varphi \, dp = \int \varphi \, dm \leq \int \varphi \, dp_x .$$

Hence p_x dominates p and the proof is complete. □

Remarks.

1. We have seen in proposition 1.5.3 that every simplex is a Choquet simplex. The examples indicated, that the converse can fail for various reasons. In fact, considerable difficulties have to be overcome to find conditions under which the converse can be proved. We note three important cases where the equivalence

(*) K is a Choquet simplex if and only if it is a simplex

holds true. The first one is the celebrated Choquet-Meyer uniqueness theorem (Phelps(1966)):

(*) holds if K is a compact convex subset of a locally convex space.

The second one reads

(*) holds if K is a subset of a Banach space which is closed bounded convex and has the Radon-Nikodym Property.

This was proved in Bourgin/Edgar(1976), thm. 4.3. It seems still to be an open problem if the Radon-Nikodym Property can be dropped in the proof of the statement "if K is a Choquet simplex then every element of K is the barycenter of *at most* one boundary measure" (Edgar(1978),

page 74). For further information, we refer to Bourgin(1983).
The last example is proposition 1 in v. Weizsäcker/Winkler(1980):

(*) holds if K is a measure convex subset of a locally convex space and every set of discrete probability measures on K which is directed upwards in Choquet order has an upper bound in Choquet order.

2. For applications, results of the following type are convenient:

(#) K is a Choquet simplex if and only if every element of K is the barycenter of a unique tight probability measure on the extreme boundary of K.

In view of proposition 1.4.2, the Choquet-Meyer theorem becomes:

(#) holds if K is a compact convex metrizable subset of a locally convex space.

The Bourgin/Edgar theorem for separable sets corresponding to that noted above (Bourgin/Edgar(1976), thm. 1.1) was generalized by E. Thomas ((1980), cor. 1 on page 504):

(#) holds if K is a closed Suslin measure convex subset of a locally convex space and K has the Radon-Nikodym Property.

Finally, we note a result from v. Weizsäcker/Winkler(1980):

(#) holds if K is a Suslin measure convex subset of a locally convex space and K enjoys the martingale condition:
every K-valued martingale $(f_i, F_i)_{i \in \mathbb{N}}$ on a probability space $(X, F, \mathbb{P})$ with constant f_1 can be extended to a K-valued martingale $(f_i, F_i)_{i \in \mathbb{N} \cup \{\infty\}}$ on $(X, F, \mathbb{P})$.

Here, martingales are defined along the same lines as conditional expectations in definition 1.3.5.

Because of its importance, we describe in a few words the development of the notion "Choquet simplex".

Historical note concerning Choquet simplices.
Let us call a nonvoid convex subset of a linear space H a base iff its minimal affine extension $\{\alpha x + (1-\alpha)y: \alpha \in \mathbb{R}; x,y \in B\}$ does not meet the origin. In fact, B is the base of the pointed convex cone $C := \{\alpha x: \alpha \geq 0, x \in B\}$. Any nonvoid convex set can be made a base by embedding it into $H \times \mathbb{R}$ via the mapping $x \to (x,1)$. A homothet of B is a set of the form $x + \alpha B$ with $x \in H$ and $\alpha > 0$.

G. Choquet was the first to notice that if H is Euclidean n-space and B is compact, then B is a finite dimensional simplex if and only if the following condition holds:

(Σ) whenever the intersection of two homothets of B is at least one-dimensional, then it is itself a homothet of B.

Choquet used this condition to define simplices in infinite dimension (Choquet(1956a), (1956b)).

Independently of Choquet, C.A. Rogers and G.C. Shephard (1957) showed that among the finite-dimensional compact bases the simplices are exactly those sets B which have the property

(Σ') whenever the intersection of two translates of B is at least one-dimensional then the intersection is a homothet of B.

In fact, (Σ') can also be formulated in infinite dimension and is equivalent to (Σ) (Eggleston/Grünbaum/Klee(1964)).

G. Choquet formulated two further conditions (C) and (L):

(C) for $x,y \in C$ there is $z \in C$ such that $(x+C) \cap (y+C) = z+C$,

(L) C is a lattice cone in its own order.

Obviously, (C) and (L) are equivalent.

The conditions (Σ), (Σ'), (C) and (L) are equivalent as soon as B is linearly compact, i.e. if every line intersects B in a set which is compact in the Euclidean topology of the line (Kendall(1962)). Choquet formulated (L) simultaneously with (Σ) to state his theorem on existence and uniqueness of integral representation in compact convex sets (Choquet(1956a), (1956b)). Results of that type are usually formulated in terms of condition (L) and convex sets with property (L) are referred to as Choquet simplices. A proof that a compact base in $\mathbb{R}^n$ is a finite-dimensional simplex if and only if (L) holds is given in Phelps(1966), pages 75-76.

It should be mentioned that the connection between simplices and vector lattices was observed as early as 1947 by J.A. Clarkson (1947).

CHAPTER 2

FOUR ASPECTS OF CHOQUET ORDER

Again measures on measure convex sets are considered. We prove four results for measures on such sets which are related in Choquet order.

In section 2.1, it is shown that any measure on a locally convex space dominated in Choquet order by a measure on a measure convex subset also lives on that subset.

In section 2.2, we provide a criterion under which sets of measures bounded from above are uniformly tight in a strict sense. This criterion is fulfilled, if the upper bound is concentrated on a measure convex set.

In section 2.3, it is studied, when a measure on the weak Borel σ-algebra of a locally convex space can be extended to the strong Borel σ-algebra. The connection with Choquet order is established.

In section 2.4, it is proved that on measure convex sets nets of measures decreasing in Choquet order have a greatest lower bound and converge to this infimum. An analogue for increasing nets holds too. The former is the crucial tool in deriving the main results of chapter 3.

If nothing else is stated, the results were obtained by the author.

Section 2.1. Measures smaller in Choquet order live on smaller sets

The headline roughly expresses, what one naively would expect from Choquet order. Here again, measure convexity plays an essential role, as proposition 2.1.3 will show.

Before, a simple measure theoretic observation:

2.1.1 Lemma. Let (X,t) be a σ-compact topological space and s a topology on X coarser than t. Then:

(i) $\mathcal{B}(X,s) = \mathcal{B}(X,t)$,

(ii) $P(X,s) = P(X,t)$.

Proof. Choose an increasing sequence of subsets C_i, $i \in \mathbb{N}$, which are t-compact such that $X = \bigcup_{i\in\mathbb{N}} C_i$. Recall that for these compact sets $C_i \cap s = C_i \cap t$.

For a family G of subsets of X denote by $\sigma(G)$ the σ-algebra on X generated by G. Then

$$\mathcal{B}(X,t) = \sigma\Big(\bigcup_{i\in\mathbb{N}} (C_i \cap t)\Big) = \sigma\Big(\bigcup_{i\in\mathbb{N}} (C_i \cap s)\Big) = \mathcal{B}(X,s)$$

which proves (i) while (ii) follows from

$$P(C_i, C_i \cap t) = P(C_i, C_i \cap s) \quad \text{for every } i \in \mathbb{N}. \qquad \square$$

We will need the more technical version 2.1.2 of proposition 2.1.3. The set in question is not measure convex in the original locally convex topology, but only in a stronger one.

2.1.2 Lemma. Let F be a linear space with a locally convex topology σ and E a linear subspace with a locally convex topology τ stronger than $\sigma \cap E$ and denote by ι the embedding map of E into F. Assume further that $E' = \{l|E : l \in F'\}$.

Consider measures $\tilde{p} \in P'(F,s)$ and $q \in P'(E,t)$ such that

$$\tilde{p} < \tilde{q} := q \circ \iota^{-1}.$$

If M is a measure convex subset of E then $q_*(M) = 1$ implies $\tilde{p}_*(M) = 1$.

Proof.

1. Assume for the present that $\tilde{p}(\tilde{C}) = 1$ for some subset $\tilde{C}$ of F which is compact in σ. Choose a subset C of M, σ-compact in τ with $q(C) = 1$. There is a set B in F bounded in σ containing both, $\tilde{C}$ and C. Now the equivalence a.⇔e. from theorem 1.3.6 applies; for $\tilde{p}$ and $\tilde{q}$ there is $\mu \in P(P(B, B \cap \sigma))$ with respective properties e.(i) and (ii).
In particular

$$1 = \tilde{q}(C) = \int_{P(B)} \tilde{m}(C) \, d\mu(\tilde{m});$$

so for the set

$$A := \{\tilde{m} \in P(B, B \cap s) : \tilde{m}(C) = 1\}$$

one has

$$\mu(A) = 1.$$

All measures from A live on C, so by lemma 2.1.1 they may be identified with elements from $P(C, C \cap \tau)$ and consequently, they have their barycenter in M (recall the assumption about E'!):

$$r[A] \subset M.$$

Since μ is tight, there is a σ-compact subset A_μ of A with $\mu(A_\mu) = 1$; proposition 1.1.3 assures that the barycenter map is continuous, so by 1.3.6.e.(ii)

$$\tilde{p}_*(M) \geq \tilde{p}(r[A_\mu]) = \mu(A_\mu) = 1.$$

2. Let now $\tilde{p}$ be arbitrary. Choose pairwise disjoint in σ compact sets $\tilde{C}_i$ such that $\tilde{p}(\tilde{C}_i) > 0$ and $\tilde{p}(\bigcup_{i \in \mathbb{N}} \tilde{C}_i) = 1$. There is a $\tilde{p}$-dilation T with $\tilde{q} = \tilde{p}T$ by 1.3.6; set

$$\tilde{q}_i(B) := \int 1_{C_i} T_x(B) \, d\tilde{p}(x) \text{ for every } B \in \mathcal{B}(F, \sigma).$$

Since $\tilde{q} = \sum_{i \in \mathbb{N}} \tilde{q}_i$, each $\tilde{q}_i$ is concentrated on M; T is also a $\tilde{p}(\cdot \cap C_i)$-dilation (to be formally correct, we had to normalize!), so according to point 1, $\tilde{p}(\cdot \cap C_i)$ is concentrated on M, hence also $\tilde{p}$ itself. □

The announced result is an easy consequence.

<u>2.1.3</u> <u>Proposition</u>. Let E be a complete locally convex space. Consider a bounded subset of E. Then the following are equivalent:

a. M is measure convex;

b. whenever p, $q \in P'(E)$ such that $p < q$, then $q_*(M) = 1$ implies that $p_*(M) = 1$.

<u>Proof</u>. Assume that a. holds. Set E = F, $\sigma = \tau$ and $\tilde{p} = p$ in the preceding lemma to get a.
To prove the converse, assume that M is not measure convex but bounded. In view of proposition 1.1.3 there is some $q \in P(M)$ with $r(q) \in E \setminus M$. For $p := \varepsilon_{r(q)}$, we have obviously $p < q$, but even $p^*(M) = 0$. □

<u>Section 2.2. Uniform tightness of sets of measures bounded from above in Choquet order; convex tight measures</u>

A further pleasant property of Choquet order is revealed: sets of probability measures bounded above are uniformly tight in a strict sense, if the upper bound is concentrated on a measure convex set (corollary 2.2.4). The salient point is a regularity condition of measures on measure convex sets.

<u>2.2.1</u> <u>Definition</u>. Let M be a subset of a locally convex space E. $p \in P(M)$ is called <u>convex-tight</u> iff

$$\sup\{p(C) : C \subset M \text{ convex and compact}\} = 1.$$

A subset $\mathcal{Q}$ of $P(M)$ is said to be uniformly convex-tight in M' where $M \subset M' \subset E$ iff for every $\varepsilon > 0$ there is a convex compact subset C of M' such that

$$p(M' \setminus C) < \varepsilon \quad \text{for every } p \in \mathcal{Q} .$$

If $\mathcal{Q}$ is uniformly convex tight in M' this may fail in M!

Example 1. Denote by M the set of irrational numbers in $M' := [0,1]$. The tightness condition is trivially fulfilled for each $p \in P(M)$ in M':

$$\sup\{p(C) : C \subset [0,1] \text{ convex and compact}\} = 1 .$$

But obviously

$$\sup\{p(C) : C \subset M \text{ convex and compact}\} = 0 .$$

The proof of the following result contains the main arguments.

2.2.2 Proposition. Suppose that M is a bounded subset of the complete locally convex space E. Assume further that a fixed measure $p \in P(M)$ is convex-tight. Then the set

$$\mathcal{Q} := \{p \in P(M) : p \prec q\}$$

is uniformly convex-tight in E.

Proof. Let Γ denote the family of all continuous semi-norms γ on E satisfying

$$\sup\{\gamma(x-y) : x, y \in M\} \le 1 .$$

Γ generates the topology of E. Choose now $\varepsilon > 0$.

1. By assumption, for every $i \in \mathbb{N}$ there is a compact convex subset C_i of M such that

$$(1) \qquad q(M \setminus C_i) \le 2^{-2i}\varepsilon .$$

Fix now some $\gamma \in \Gamma$. The function

$$\varphi_i^\gamma : E \to \mathbb{R}_+ , \ x \to \min\{\gamma(x-y) \ : y \in C_i\}$$

is continuous and - C_i being convex - also convex; further

(2) $$\varphi_i^\gamma | M \leq 1 .$$

Let B_γ denote the closed unit ball w.r.t. γ. The set

(3) $$C_i^\gamma := \{x \in E : \varphi_i^\gamma(x) \leq 2^{-i}\} = C_i + 2^{-i} B_\gamma$$

is convex and closed.

Let us show, that

$$C^\gamma := \bigcap_{i \in \mathbb{N}} C_i^\gamma$$

is totally bounded. For $\delta > 0$ choose $j > 0$ such that $2^{-i} \leq \frac{\delta}{2}$ for every $i \geq j$. Then

$$C^\gamma \subset C_j + 2^{-j} B_\gamma \quad \text{and } C_j \subset F + \frac{\delta}{2} B_\gamma$$

for some finite subset F of E. Hence

$$C^\gamma \subset F + \frac{\delta}{2} B_\gamma + 2^{-j} B_\gamma \subset F + \delta B_\gamma .$$

In summary, C^γ is totally bounded w.r.t. γ, closed and convex.

2. Choose any p from $\mathcal{Q}$. Recall that by proposition 1.3.2 we have $\int \varphi \, dp \leq \int \varphi \, dq$ for every bounded convex continuous function φ on M. Combining this with the estimates (1) and (2), $q \succ p$ and (3) yields

$$2^{-2i}\varepsilon \geq q(M \setminus C_i) \geq \int_M \varphi_i^\gamma \, dq \geq \int_M \varphi_i^\gamma \, dp \geq 2^{-i} p(M \setminus C_i^\gamma) .$$

Hence

$$p(M \setminus C_i^\gamma) \leq 2^{-i}\varepsilon$$

and in consequence

(4) $p(C^{\gamma}) \geq 1 - \varepsilon.$

3. From part 1 we see that the set

$$C := \bigcap_{\gamma \in \Gamma} C^{\gamma}$$

is convex closed and totally bounded w.r.t. any γ in Γ, hence compact in E (Robertson/Robertson(1967), V.4.15).

The family Γ contains the maxima of any two of its members; thus the family of sets $(C^{\gamma})_{\gamma \in \Gamma}$ is directed downwards. Since p is τ-smooth, the estimate (4) implies that

$$p(C) = \inf\{p(C^{\gamma}) : \gamma \in \Gamma\} \geq 1 - \varepsilon.$$

This completes the proof. □

On measure convex sets, the hypothesis on the upper bound of $\mathcal{Q}$ is always fulfilled.

2.2.3 Proposition. If M is measure convex then every $q \in P(M)$ is convex-tight.

Proof. Since q is tight, we have

$$1 = \sup\{q(C) : C \subset M \text{ compact}\}.$$

By proposition 1.2.5 the closed convex hull of each compact subset of M is contained in M and compact. So q is convex-tight. □

For measure convex sets an improved version of proposition 2.2.2 holds.

2.2.4 Corollary. Suppose that M is a measure convex subset of the complete locally convex space E. Assume that $q \in P(E)$ satisfies $q_*(M) = 1$. Then every element p of

$$\mathcal{Q}' := \{p \in \mathcal{P}'(E) : p < q\}$$

also satisfies $p_*(M) = 1$ and $\mathcal{Q}'$ is uniformly convex-tight in E.

Proof. By proposition 2.1.3 every $p \in \mathcal{Q}'$ is concentrated on M, i.e. $p_*(M) = 1$. So in view of proposition 2.2.3, the assertion can be read off from proposition 2.2.2. □

Because of their importance, we note a version for complete bounded and convex sets. Completeness of the underlying locally convex space may be dropped and the uniform convex-tightness is stricter.

2.2.5 Corollary. Suppose that M is a complete bounded and convex subset of a locally convex space E. Assume that $q \in \mathcal{P}(E)$ satisfies $q(M) = 1$. Then every element p of

$$\mathcal{Q}' := \{p \in \mathcal{P}'(E) : p < q\}$$

satisfies $p(M) = 1$ and $\mathcal{Q}'$ is uniformly convex-tight in M.

Proof. The set M is measure convex by corollary 1.2.4. So from corollary 2.2.4 follows that $p(M) = 1$ and $\mathcal{Q}'$ is uniformly convex-tight in the completion $\tilde{E}$ of E. Since for every compact convex subset $\tilde{C}$ of $\tilde{E}$ the intersection with M is compact and convex, the assertion is proved. □

Remark. Convex-tight measures play an important role in the theory of "large deviations" and related fields. From the immense amount of literature, we only mention Bahadur/Zabell(1979) and Csiszár(1984). Another important property is convex-regularity. A probability measure $p \in P(E)$ is convex-regular iff $p(G) = \sup\{p(C) : C \subset G, C \text{ convex and compact}\}$ for every open convex subset G of E. By proposition 1 in the first reference, p. 619, a measure $p \in \mathcal{P}(E)$ is convex-tight if and only if it is convex-regular. Hence by proposition 2.2.3 above, each tight probability measure on a measure convex set is convex-regular.

Section 2.3. Extension of tight measures from the weak Borel σ-algebra to the strong Borel σ-algebra

Let E be a locally convex space with topology τ. We call $\mathcal{B}(E,\sigma(E,E'))$ the _weak_ and $\mathcal{B}(E,\tau)$ the _strong_ Borel σ-algebra of E. We consider the question when a "weak" measure $p' \in \mathcal{P}(E,\sigma(E,E'))$ has a "strong" extension $p \in \mathcal{P}(E,\tau)$.

If (E,τ) is metrizable, each weak measure has a strong extension according to a well-known theorem due to Phillips, Dunford-Pettis and Grothendieck (see proposition 2.3.4 below). A rather lengthy proof is given in Schwartz(1973), pages 162-166. We give a short proof which was indicated to us by J.P.R. Christensen (1983b). It is based on a theorem of R.E. Johnson (1969) which was generalized and supplied with a simpler proof by Christensen ((1983a), theorem 2). As far as I know, there is no example im literature which shows that the statement of proposition 2.3.4 may fail to be true without the metrizability assumption. A counterexample is supplied in this section. This material was published in Winkler(1984a).

What has all that to do with Choquet order? Indeed a lot:
On measure convex sets the conclusion of the Phillips-Dunford-Pettis-Grothendieck theorem holds if and only if the weak measure is dominated in Choquet order by a strong measure. This will be shown in proposition 2.3.5.

The following simple lemma settles the uniqueness question.

_2.3.1_ _Lemma_. Let X be a set with two topologies t and t', where t is finer than t'.
If $p' \in \mathcal{P}(X,t')$ has an extension $p \in \mathcal{P}(X,t)$ then

$$\mathcal{B}(X,t) \subset \mathcal{B}_{p'}(X,t').$$

In particular, the extension is unique.

<u>Proof</u>. Consider some $B \in \mathcal{B}(X,t)$. There are subsets C of B and D of $X \setminus B$ which are σ-compact for the topology t and satisfy

$$p(B \setminus C) = 0 = p((X \setminus B) \setminus D).$$

Because C and D are also Borel sets w.r.t. the weaker topology t', the lemma is proved. □

We state Johnson's theorem in a version sufficient for our needs.

<u>2.3.2</u> <u>Proposition</u>. Let X and Y be compact spaces. Assume further that X is the support of some tight probability measure. Then:
if $f : X \times Y \to \mathbb{R}$ is a separately continuous function, the set $\{f(x,\cdot): x \in X\} \subset C(Y)$ is separable in the supremum norm.

<u>Proof</u>. Christensen(1983a) □

The essential step in the proof of proposition 2.3.4 is the following result.

<u>2.3.3</u> <u>Proposition</u>. Let E be a metrizable locally convex space with metric topology τ. Assume further that $p \in P(E,\sigma(E,E'))$ has weakly compact support C. Then the following holds:

a. the weak and the strong Borel σ-algebras coincide on C;

b. the topological space $(C, C \cap \tau)$ is Polish; in particular the restriction of p to C is in $P(C, C \cap \tau)$.

<u>Proof</u>.

1. Assume for the present that E is a Banach space. Denote by B' the closed unit ball of E'; it is compact in the weak*-topology $\sigma(E',E)$. Apply Johnson's theorem 2.3.2 to the evaluation map

$$f : C \times B' \to \mathbb{R}, \quad (x,\varphi) \to \varphi(x)$$

to conclude that $\{f(x,\cdot): x \in C\} \subset C(B')$ is separable in the supremum norm. The mapping

$$C \ni x \to f(x,\cdot) \in C(B')$$

is an isometry, hence C itself is norm separable.

2. Assume now that E is metrizable. There are Banach spaces E_i, $i \in \mathbb{N}$, and an isomorphism ι of E into $\prod_{i=1}^{\infty} E_i$ (Bessaga/Pelczyński(1975), thm. I.6.3). Denote the projections by pr_i and set $\iota_i := pr_i \circ \iota$. Then $p \circ \iota_i^{-1} \in P(E_i, \sigma(E_i, E_i'))$ and the support $\iota_i(C)$ is weakly compact, hence separable in norm by part 1. It follows that C is separable in metric since

$$\iota(C) \subset \prod_{i=1}^{\infty} \iota_i(C).$$

Furthermore C being weakly complete, it is complete in the metric topology (Köthe(1969), 18.4(4), page 210).

Putting things together, we get that C is Polish in the metric topology. Thus the weak and the strong Borel σ-algebra on C coincide (Schwartz (1973), cor. 2 on page 101). On a Polish space every probability measure is tight. This concludes the proof. □

The extension result announced in the introduction reads as follows.

2.3.4 Proposition. Let E be a metrizable locally convex space with metric topology τ. Then every $p' \in P(E, \sigma(E,E'))$ has a unique extension $p \in P(E,\tau)$.

Proof. Observe that p' is concentrated on a countable union of pairwise disjoint weakly compact sets. Apply proposition 2.3.3 and lemma 2.3.1 to get the conclusion. □

The following example shows that proposition 2.3.4 cannot be extended to arbitrary locally convex spaces. The same construction can be carried out in a large class of locally convex spaces.

Example. Let I be an uncountable index set and for each $i \in I$ let E_i be a copy of $\ell^2(\mathbb{N})$; denote the norm toplogy by τ_i. Let further denote E the product of these spaces and τ the product topology. We construct a measure $p' \in P(E,\sigma(E,E'))$ which has no extension $p \in P(E,\tau)$. First of all, fix a measure $\mu \in P(\ell^2(\mathbb{N}))$ which is concentrated on a weakly compact subset C of $\ell^2(\mathbb{N})$ and fulfills $\mu(K) < 1$ for every strongly compact subset K of $\ell^2(\mathbb{N})$; take for example

$$\mu = \sum_{n\in\mathbb{N}} 2^{-n}\varepsilon_{e_n},$$

where e_n denotes the n-th unit vector of $\ell^2(\mathbb{N})$.

For copies μ_i of μ the product measure on $\bigotimes_{i\in I} \mathcal{B}(E_i,\sigma(E_i,E_i'))$ has a unique tight extension p' to $\mathcal{B}(E, \prod_{i\in I} \sigma(E_i,E_i'))$ by Kakutani(1943), thm. 2 (see also Dalgas(1982), 3.16). Because

$$\sigma(E,E') = \prod_{i\in I} \sigma(E_i,E_i')$$

(Schaefer(1971),IV.4.3), we have constructed a measure

$$p' \in P(E,\sigma(E,E')).$$

We will show now that $p'(K) = 0$ for every strongly compact subset K of E. Consequently, p' has no tight extension to the strong Borel σ-algebra of E.

For finite subsets J of I denote by

$$pr_J : E \to \prod_{j\in J} E_j$$

the canonical projections and by

μ_J the product measure on $\mathcal{B}(\prod_{j\in J} E_j, \prod_{j\in J} \sigma(E_j,E_j'))$.

According to the choice of μ, we have

$$\mu_i(pr_{\{i\}}[K]) < 1 \text{ for every } i \in I.$$

Since I is uncountable, at least countably many of these quantities are bounded away from 1. This implies

$$\inf\left\{\prod_{j\in J} \mu_j(\mathrm{pr}_{\{j\}}[K]) \ : \ J \text{ finite subset of } I\right\} = 0.$$

As

$$K \subset \bigcap_{\substack{J\subset I\\ \text{finite}}} \mathrm{pr}_J^{-1}\left[\mathrm{pr}_J[K]\right],$$

we get

$$\begin{aligned}
&p'(K) \leq \\
&\leq \inf\left\{p'\left(\mathrm{pr}_J^{-1}\left[\mathrm{pr}_J[K]\right]\right) \ : \ J\subset I \ \text{ finite}\right\} = \\
&= \inf\left\{\mu_J(\mathrm{pr}_J[K]) \ : \ J\subset I \ \text{ finite}\right\} \leq \\
&\leq \inf\left\{\prod_{j\in J} \mu_j(\mathrm{pr}_{\{j\}}[K]) \ : \ J\subset I \ \text{ finite}\right\} = \\
&= 0.
\end{aligned}$$

This completes the example.

I thank Z. Lipecki for an improvement of proposition 2.3.3 which led to a simplification of the proof for 2.3.4. He also reminded me to mention the following.

There are spaces which are not metrizable but for which the conclusion of proposition 2.3.4 holds. Trivially it is true for Spaces of Minimal Type - i.e. spaces which admit no strictly coarser separated locally convex topology. They are exactly the locally convex spaces isomorphic to $\mathbb{R}^I$ (Schaefer(1971), ex. IV.6, page 191). This leads to the

Problem. Characterize all locally convex spaces for which the conclusion of proposition 2.3.4 holds.

We state now the announced extension result used in the sequel.

<u>2.3.5</u> <u>Proposition</u>. Let E be a locally convex space with topology τ and M a measure convex subset of E. Consider measures $p' \in P(M, M\cap\sigma(E,E'))$ and $q \in P(M, M\cap\tau)$ such that

$$p' < q' := q|\mathcal{B}(M, M\cap\sigma(E,E')).$$

Then there is a unique measure $p \in P(M, M\cap\tau)$ with $p < q$ extending p'.

<u>Proof</u>. By the lemma in 0.7 a tight probability measure on M is uniquely determined by its values on $S(M)$. So it is sufficient to construct $p \in P(M, M\cap\tau)$ which is dominated in Choquet order by q and coincides with p' on $Z(M)$ - the uniqueness assertion follows from lemma 2.3.1.

1. We use the characterization of Choquet order from proposition 1.3.6.c.

There are a probability space $(X, F, \mathbb{P})$ and a sub-σ-algebra G of F and mappings g' and f' from X into M such that

(i) g' is G-$\mathcal{B}(M, M\cap\sigma(E,E'))$ measurable and $p' = \mathbb{P}\circ(g')^{-1}$,

(ii) f' is F-$\mathcal{B}(M, M\cap\sigma(E,E'))$ measurable and $q' = \mathbb{P}\circ(f')^{-1}$,

(iii) $g' = \mathbb{E}(f'|G)$.

We may assume that F and G are complete.

2. Because F is complete and the strong Borel σ-algebra is in the q'-completion of the weak one by lemma 2.3.1, f' is also F-$\mathcal{B}(M, M\cap\tau)$ measurable. We denote it by f if it is considered this way. Again by lemma 2.3.1, the image measure $\mathbb{P}\circ f^{-1}$ is q. Since M is measure convex, proposition A.3 ensures the existence of the conditional expectation $g = \mathbb{E}(f|G)$ with tight image measure $p := \mathbb{P}\circ g^{-1} \in P(M, M\cap\tau)$. Clearly, we have for each $G \in G$ and $l \in E'$:

$$\int_G l\circ g \, d\mathbb{P} = \int_G l\circ f \, d\mathbb{P} = \int_G l\circ f' \, d\mathbb{P} = \int_G l\circ g' \, d\mathbb{P}.$$

This shows that $l\circ g = l\circ g'$ almost everywhere. Hence for every set B of the form $B = \{x \in M : (l(x))_{l \in F} \in B^*\}$, where F is a finite subset of E' and $B^* \in \mathcal{B}(\mathbb{R}^F)$ the following holds:

(1) $$p(B) = \mathbb{P}(g^{-1}(B)) = \mathbb{P}((g')^{-1}(B)) = p'(B).$$

Condition (1) holds for every $B \in Z(M)$. Now we have constructed a measure p with the properties described above, which completes the proof.
□

Section 2.4. Nets of measures monotone in Choquet order

On measure convex sets, nets of measures decreasing in Choquet order have greatest lower bounds and converge to these in the weak topology on measures (theorem 2.4.2 below). An analogue for increasing nets will also be proved.

Such monotone nets are of central interest. Let us first indicate an aspect which will not be pursued in this paper. There is an intimate connection between convergence to the least upper bound and martingale convergence. The equivalence a.⇔c. in theorem 1.3.6 suggests, that under reasonable conditions an adapted process (f_i) should be a martingale if and only if $(\mathbb{P} \circ f_i^{-1})$ is an isotone net in Choquet order. Such a connection was revealed by J.L. Doob (1968). Further, the martingale convergence theorem holds if and only if $(\mathbb{P} \circ f_i^{-1})$ converges. In nearly all proofs of Choquet integral representation theorems such arguments are involved. The reader is referred to Bourgin(1983) and the references there, Edgar(1978), Pestman(1981), v. Weizsäcker(1977) and (1978), v. Weizsäcker/Winkler(1980). A few further remarks on that topic conclude this section.

In the present context, the version for decreasing nets will be an important tool to show that the decreasing intersection - or more generally the inverse limit - of simplices is again a simplex.

We give first a simple proof for the compact case.

2.4.1 Lemma. Let M be a compact subset of a locally convex space. Consider a net $(p_i)_{i \in I}$ in $P(M)$ decreasing (increasing) in Choquet order. Then there is a measure $p \in P(M)$ such that

(i) p is the greatest lower (least upper) bound of $(p_i)_{i \in I}$ in Choquet order,

(ii) $p_i \overset{w}{\to} p$.

Proof. We only write down the proof for decreasing nets. For increasing nets it is - mutatis mutandis - the same.

1. In view of the Riesz representation theorem the set $P(M)$ is compact in $\sigma(M(M), C(M))$. Thus the net $(p_i)_{i \in I}$ has a cluster point p. For every $\varphi \in S(M)$ the real number $p(\varphi)$ is a cluster point of the decreasing net $(p_i(\varphi))_{i \in I}$, hence its limit:

$$\int_M \varphi \, dp_i \downarrow \int_M \varphi \, dp \quad \text{for every} \quad \varphi \in S(M).$$

2. It follows that

$$\int_M \varphi \, dp_i \to \int_M \varphi \, dp \quad \text{for every} \quad \varphi \in S(M) - S(M).$$

By the lattice version of the Stone-Weierstraß theorem (cf. 0.7) the space $S(M) - S(M)$ is uniformly dense in $C(M)$, from which we conclude that $p_i \overset{w}{\to} p$. That p is the greatest lower bound w.r.t. Choquet order follows immediately from part 1 of this proof. □

Now, we formulate the central result of this section.

2.4.2 Theorem. Let M be a measure convex subset of some locally convex space. Consider a net $(p_i)_{i \in I}$ in $P(M)$. Then:

a. If $(p_i)_{i \in I}$ decreases in Choquet order, there is a greatest lower bound p in Choquet order and $p_i \overset{w}{\to} p$.

b. If $(p_i)_{i \in I}$ increases and has on M an upper bound in Choquet order, then there is a least upper bound p in Choquet order and $p_i \overset{w}{\to} p$.

Proof. Let us introduce some notation. Denote the underlying locally convex space by E and its topology by τ, the space E'* by F and the topology $\sigma(E'^*,E')$ by σ; finally write ι for the embedding map $E \hookrightarrow E'^*$. Note that the closure $\tilde{M}$ of $\iota[M]$ in σ is compact.

1. Consider a decreasing net $(p_i)_{i\in I}$. It is obvious by the characterization 1.3.6.a of Choquet order that the net $(p_i \circ \iota^{-1})_{i\in I}$ also decreases in Choquet order. By lemma 2.4.1, it has a greatest lower bound $\tilde{p} \in P(\tilde{M})$. Lemma 2.1.2 ensures that $\tilde{p}_*(\iota[M]) = 1$. The spaces

$$(\iota[M], \iota[M] \cap \sigma) \quad \text{and} \quad (M, M \cap \sigma(E,E'))$$

are affinely homeomorphic, thus $\tilde{p}\circ\iota$ is the greatest lower bound in $P(M, M\cap\sigma(E,E'))$ of the net $(p_i | B(M, M\cap\sigma(E,E')))_{i\in I}$. Proposition 2.3.5 allows to extend $\tilde{p}\circ\iota$ to the tight probability measure p on the strong Borel σ-algebra $B(M, M\cap\tau)$ which is the greatest lower bound of the net $(p_i)_{i\in I}$.

To construct the least upper bound for an increasing net, we proceed in complete analogy up to the fact that the least upper bound of $(p_i \circ \iota^{-1})_{i\in I}$ is concentrated on $\iota[M]$. But this holds again by lemma 2.1.2 because the original net is bounded above by assumption.

Note that in either case

$$(1) \qquad \int_M \varphi \, dp_i \to \int_M \varphi \, dp \quad \text{for every} \quad \varphi \in S(M) - S(M).$$

2. We turn to the assertion "$p_i \overset{w}{\to} p$" for decreasing as well as increasing nets.

To begin with, assume that E is complete. Choose any continuous function $f : M \to [-1,1]$. Choose further a positive number ε and some $j \in I$. By corollary 2.2.4 there is a compact subset C of E such that

$$(2) \qquad p(M\setminus C) < \varepsilon \quad \text{and} \quad p_i(M\setminus C) < \varepsilon \quad \text{for every} \quad i \geq j.$$

In view of the lemma in 0.7, the lattice version of the Stone-Weierstraß theorem applies to the linear space $S(C) - S(C)$; thus it is uniformly dense in $C(C)$. There is an element $\psi \in S(M) - S(M)$ such that

(3) $\sup\{|\psi(x) - f(x)| : x \in C\} < \varepsilon.$

Since $S(M) - S(M)$ is a vector lattice, ψ may be assumed to fulfil

(4) $\sup\{|\psi(x)| : x \in M\} \leq 1.$

Finally, according to (1) there is $k \geq j$ such that

(5) $|\int_M \psi \, dp_i - \int_M \psi \, dp| < \varepsilon$ for every $i \geq k$.

Together, the inequalities (2) - (5) yield for every $i \geq k$ the estimate

$$\begin{aligned} &|\int_M f \, dp_i - \int_M f \, dp| \leq \\ &\leq |\int_C (f-\psi) \, dp_i| + |\int_C (f-\psi) \, dp| + |\int_M \psi \, dp_i - \int_M \psi \, dp| + \\ &+ 2\cdot(\sup\{|f(x)| : x \in M\} + \sup\{|\psi(x)| : x \in M\})\cdot\varepsilon < \\ &< 7\cdot\varepsilon. \end{aligned}$$

In consequence,

$$\int f \, dp_i \to \int f \, dp \quad \text{for each} \quad f \in C_b(M)$$

which proves the assertion for complete spaces E. As in this formula only functions on M are involved, we may consider M as a subset of the completion if E is not complete itself. Hence the assertion holds for every locally convex space. □

Remark. If E is metrizable, in the proof of the existence of greatest lower or least upper bounds the classical theorem formulated here as proposition 2.3.4 can be substituted for proposition 2.3.5 (the assertion that $p < q$ in 2.3.5 is trivial).

In an important special case of theorem 2.4.2, the proof can be considerably simplified - namely when E is endowed with the weak topology. Before, we note a very useful lemma which goes back to L. Le Cam

((1957), page 216). An algebra version is proved in Hoffmann-Jørgensen ((1975), lemma 2.1 on page 76) and ((1977), lemma 4.1 on page 27). We need the lattice version.

2.4.3 Lemma. Let (X,t) be a completely regular space. Suppose that G is a subset of $C_b(X,t)$ such that

(i) $f \vee g \in G$ whenever $f,g \in G$,

(ii) G contains the constant functions on X,

(iii) G separates points on X.

Denote by t' the topology on X generated by G. If $(\mu_i)_{i\in I}$ is a net in $P(X,t')$ and $\mu \in P(X,t')$ with

$$\mu_i(g) \to \mu(g) \quad \text{for every} \quad g \in G$$

then $\mu_i \overset{w}{\to} \mu$ in $P(X,t')$.

Note that - in contrast to the above mentioned algebra version - condition (ii) is necessary! (Cf. Edgar(1978), proposition 2.1.) This is easily seen if one considers as G the set of all linear functions on the unit interval.

Proof. Copy the proof in Hoffmann-Jørgensen(1977), but use the lattice version of the Stone-Weierstraß theorem instead of the algebra version.
□

2.4.4 Supplement to theorem 2.4.2.
Suppose that in theorem 2.4.2 it is assumed additionally that E is endowed with the weak topology $\sigma(E,E')$. Then the proof can be simplified as follows.

Proof. To begin with, proceed as in the general case to get $\tilde{p}$. The measure $\tilde{p}\circ\iota$ needs not to be extended, so the use of proposition 2.3.5 is superfluous. One has again

(1) $\int_M \varphi \, dp_i \to \int_M \varphi \, dp$ for every $\varphi \in S(M) - S(M)$.

Since $S(M)$ generates the weak topology, the convergence $p_i \overset{w}{\to} p$ follows immediately from lemma 2.4.3. □

In theorem 2.4.2 measure convexity cannot be dropped without any substitute.

Example. Denote by e_i, $i \geq 1$, the unit vectors of $\ell^1(\mathbb{N})$ and set $E := \text{span}\{e_i : i \geq 1\}$. Set further

$$x_n := (1-2^{-n})^{-1} \left(\sum_{i \leq n} 2^{-i} e_i \right)$$

and define

$$p_n := (1-2^{-n}) \varepsilon_{x_n} + \left(\sum_{i>n} 2^{-i} \varepsilon_{e_i} \right).$$

The net $(p_n)_{n \geq 1}$ on the unit ball of E decreases in Choquet order. The unit ball B of $\ell^1(\mathbb{N})$ is measure convex; in $P(B)$, we have $p_n \overset{w}{\to} \varepsilon_x$, where

$$x = \sum_{i \geq 1} 2^{-i} e_i.$$

Hence ε_x is the greatest lower bound by theorem 2.4.2. Since $x \notin E$, there is no greatest lower bound in $P(B \cap E)$.

Increasing nets are not treated in the sequel; we confine ourselves to several remarks.

Remark. Let us quote from Edgar(1978), lemma 2.3:

Lemma. Let $(P, \leq)$ be a partially ordered set. Suppose that every chain in P has a least upper bound. Then every subset of P which is directed has a least upper bound.

Combining this lemma with theorem 2.4.2.b yields

Proposition. Let M be a measure convex set. Suppose that every increasing chain in $(P(M), <)$ has an upper bound. Then each increasing net in $(P(M),<)$ has a least upper bound and converges to it in $\sigma(M(M),C_b(M))$.

This proves and improves - at least for measure convex sets - the implication (3)⇒(2) of theorem 2.2 in v. Weizsäcker(1978); I was not able to follow the arguments there.

Theorem 2.4.2.b also shows that on measure convex sets the condition

3) $p_i(\varphi) \uparrow p(\varphi)$ for every $\varphi \in S(M)$

implies

1) $p_i(f) \to p(f)$ for every $f \in C_b(M)$,

as claimed in Satz 20 on page 74 of v. Weizsäcker(1977). This proof is direct; a detour using results from Doob(1968) is not necessary. The latter is not easy, since Doob's result holds only for chains of measures and not for arbitrary nets. In v. Weizsäcker's proof there is a gap.

CHAPTER 3

INVERSE LIMITS OF SIMPLICES

The following question suggests itself: "Under which conditions is the inverse limit of an inverse system of simplices again a simplex?". It is of interest both in abstract theory and in view of applications. This question will be answered completely in this chapter. The proofs illustrate how to use the techniques developed in chapter 2. The results were obtained by the author (cf. Winkler(1983)).

A variety of authors answered the question on different levels of generality in the case of compact simplices (cf. the historical note at the end of this chapter). If the simplices are all compact then the inverse limit is itself a convex and compact set; hence Choquet's theorem applies and the problem of *existence* of extremal integral representation is settled from the beginning. Only *uniqueness* of representing boundary measures remains to be shown.

If the simplices are not assumed to be compact, existence *and* uniqueness of representing boundary measures have to be treated. The tools developed in chapter 2 and summarized in theorem 2.4.2 allow to answer the above question in the fairly general case of measure compact simplices as introduced in chapter 1.

This will be done in section 3.2. Before, the special case of intersections is considered. The reason is that the proofs are more simple and it can be shown more than in the case of general inverse systems (compare corollary 3.1.3 and the counterexample in section 3.2).

Section 3.1. The intersection of simplices

Existence and uniqueness of integral representation on the intersection of sets can be treated for each element of the intersection individually. The case of simplices is considered in proposition 3.1.3.

A countability condition will be needed.

3.1.1 Definition. A net $(I,\leq)$ is said to be countably generated iff there is a countable subnet $(J,\leq)$ such that for every $i \in I$ there is $j \in J$ with $i \leq j$.

The preliminary "individual" result reads:

3.1.2 Proposition. Let E be a locally convex space and $(M_i)_{i\in I}$ a net of measure convex subsets of E, decreasing w.r.t. inclusion. Assume further that every measure $q_i \in P(M_i)$ is majorized in Choquet order by a boundary measure on M_i. Consider the conditions

a. each set M_i, $i \in I$, is closed;

b. the net is countably generated.

Then each of a. or b. implies:

If an element x of the intersection $M := \bigcap_{i\in I} M_i$ is in each M_i the barycenter of a unique boundary measure p_i on M_i, then x is represented by a unique boundary measure p on M and $p_i \overset{w}{\to} p$.

Proof.

1. By assumption, for $i \leq j$ the measure p_j is dominated in Choquet order by a boundary measure on M_i, which in view of the uniqueness assumption is p_i. Hence, if we fix $k \in I$ and consider p_i, $i \geq k$, as a measure on M_k, the net $(p_i)_{i\geq k}$ in $P(M_k)$ decreases in Choquet order. By theorem 2.4.2 it converges to its greatest lower bound p, thus also the net $(p_i)_{i\in I}$ itself.

2. We prove now that $p_*(M) = 1$. By proposition 2.1.3 the measure p is concentrated on each of the measure convex sets M_i, $i \in I$.

Since every tight measure is τ-smooth, one has in case a. that

$$p(M) = \inf\{p(M_i): i \in I\} = 1.$$

From σ-continuity follows in case b. that

$$p_*(M) = \inf\{p_*(M_j): j \in J\} = 1$$

where J is countable and generates I.

3. Every $q \in P(M)$ with barycenter x is a lower bound of $(p_i)_{i\in I}$, hence

$$q(\varphi) \leq p_i(\varphi) \quad \text{for every} \quad \varphi \in S(E) \quad \text{and} \quad i \in I;$$

moreover one has $p_i(\varphi) \downarrow p(\varphi)$, which implies $q(\varphi) \leq p(\varphi)$. This shows that p is maximal and the only boundary measure on M representing x.

□

Now the corresponding result for simplices follows easily.

<u>3.1.3 Corollary</u>. Let E be a locally convex space and $(S_i)_{i\in I}$ a net of simplices in E which decreases w.r.t. inclusion. Assume further that one of the following conditions holds:

a. each simplex S_i is closed;

b. the net is countably generated.

Then the intersection $S := \bigcap_{i\in I} S_i$ is a simplex. Moreover, if $x \in S$ and p_i and p are the unique boundary measures on S_i and S with barycenter x then $p_i \overset{w}{\to} p$.

<u>Proof</u>. According to proposition 1.5.6, in a simplex every tight probability measure is dominated in Choquet order by a boundary measure. Thus we can apply proposition 3.1.2 to get existence and uniqueness of representing boundary measures on S as well as the convergence assertion. As the intersection of measure convex sets is again measure convex by lemma 1.2.6, the set S is also measure convex, hence a simplex. □

Remark. If the simplices in corollary 3.1.3 are compact, one does not need any prerequisites from chapter 2 except the simple lemma 2.4.1. This yields a short proof in this case.

Under additional assumptions, unique integral representations by means of measures *on the extreme points* are obtained.

3.1.4 Corollary. Let E be a locally convex space and $(S_i)_{i \in I}$ a net of simplices in E decreasing w.r.t. inclusion. Assume further that one of the following conditions holds:

a. each simplex S_i is closed and one of the simplices is a Suslin space;

b. the net is countably generated and for some countable subnet J of I generating I each simplex S_j, $j \in J$, is a Suslin space.

Then the intersection $S := \bigcap_{i \in I} S_i$ is a Suslin space and every element x of S is represented by a unique measure $p \in P(\text{ex } S)$.

Proof. Closed subsets of Suslin spaces as well as countable intersections of Suslin spaces are Suslin spaces (cf. Schwartz (1973), p.96). Hence the assertion follows from corollary 3.1.3 and proposition 1.4.2. □

In proposition 3.1.2 and in the corollaries 3.1.3 and 3.1.4 one neither can drop the assumption that the sets in question are closed nor that I is countably generated.

Example. Consider the measure space $([0,1], \mathcal{B}([0,1]), \lambda)$ where λ denotes Lebesgue measure as in example 2 of section 1.5. There it was noted that the set

$$TC = \{\nu \in P([0,1]) : \nu \text{ is totally continuous w.r.t. } \lambda\}$$

is no simplex (but a Suslin space).

Let N denote the collection of all Borel sets in the unit interval with

Lebesgue measure zero. For each $N \in \mathcal{N}$ the set

$$T_N := \{\nu \in P([0,1]) : \nu(N) = 0\}$$

is a simplex, since it is affinely homeomorphic to $P([0,1] \setminus N)$ (cf. Bourbaki(1969), proposition 5.8) and the latter set is a simplex (cf. example 1 in section 1.5). Moreover it is a Suslin space. So the net

$$(T_N)_{N \in \mathcal{N}}$$

is a decreasing (not countably generated) net of (in general not closed) Suslin simplices and

$$\bigcap_{N \in \mathcal{N}} T_N =$$
$$= \{\nu \in P([0,1]) : \nu(N) = 0 \text{ for every } N \in \mathcal{N}\} =$$
$$= TC;$$

so the intersection is no simplex.

Section 3.2. Inverse limits of simplices

In the last section, we considered special inverse systems of simplices. Let us turn to the general case.

3.2.1 Definition. Let $(I, \leq)$ be a directed set and E_i, $i \in I$, locally convex spaces with topologies τ_i. Let further for every $i \in I$ a convex subset L_i of E_i be given and for $i \leq j$ a continuous affine mapping $\varphi_{ji} : L_j \to L_i$.
The family of spaces $(L_i, L_i \cap \tau_i)$ together with the mappings φ_{ji} is called an inverse system (of convex subsets of locally convex spaces) iff

(i) $\varphi_{ii} = id_{L_i}$ for every $i \in I$,

(ii) $\varphi_{ji} \circ \varphi_{kj} = \varphi_{ki}$ whenever $i \leq j \leq k$.

We denote such an inverse system by $(L_i,\varphi_{ji})_{i,j\in I}$ or simply by (L_i,φ_{ji}). Let E denote the product of the spaces E_i, $i \in I$, with product topology $\tau = \prod_{i\in I} \tau_i$ and define

$$\varprojlim L_i := \{(x_i)_{i\in I} \in E: x_i \in L_i \text{ for every } i \in I,\\ \varphi_{ji}(x_j) = x_i \text{ whenever } i \leq j\}.$$

The convex set $\varprojlim L_i$ is called <u>inverse limit</u> of the system (L_i,φ_{ji}). Let finally

$$\varphi_j : \varprojlim L_i \to L_j,\ j \in I,$$

be the restrictions of the canonical projections $pr_j : E \to E_j$ on $\varprojlim L_i$.

<u>Remark</u>. In the language of category theory, a convex subset L of a locally convex space F is called inverse limit of the inverse system (L_i,φ_{ji}) if there are continuous affine mappings $\psi_i : L \to L_i$ consistent with the mappings φ_{ji} and if $\tilde{L}$ is another convex subset of some locally convex space with continuous affine mappings $\tilde{\psi}_i : \tilde{L} \to L_i$ consistent with the mappings φ_{ji}, then there is a unique affine map $\psi : \tilde{L} \to L$ such that $\tilde{\psi}_i = \psi_i \circ \psi$ for every $i \in I$.
Two such inverse limits are affinely homeomorphic (Semadeni(1971), 11.8.8(B)). So it is common to speak about *the* inverse limit. Obviously, $\varprojlim L_i$ as defined above is the inverse limit in this sense.

Let us note a simple observation.

<u>3.2.2</u>. <u>Lemma</u>. Suppose that (L_i,φ_{ji}) is an inverse system of convex sets. The set

$$\tilde{L}_k := \{(y_j)_{j\leq k} \in \prod_{j\leq k} L_j : y_j = \varphi_{kj}(y_k) \text{ for } j \leq k\}$$

is affinely homeomorphic to L_k. Further

$$\lim_{\leftarrow} L_i = \bigcap_{j\in I} \left(\tilde{L}_j \times \prod_{i\in I_j} L_i\right), \text{ where } I_j := I \setminus \{i \leq j\}.$$

Finally, $\lim_{\leftarrow} L_i$ is measure convex if each L_i is measure convex and compact if each L_i is compact.

Proof. Everything is clear except measure convexity. The latter follows from the above representation of the inverse limit and lemma 1.2.6.

□

The main result in this chapter is

3.2.3 Theorem. Let $(I,\leq)$ be an increasing net which is countably generated and $(E_i)_{i\in I}$ a family of locally convex spaces. Consider an inverse system (S_i,φ_{ji}) of simplices S_i in E_i. Then

$$S := \lim_{\leftarrow} S_i$$

is a simplex.

Proof. Fix $x = (x_i)_{i\in I} \in S$. Denote by p_i the unique boundary measure on S_i representing x_i. Define

$$p_{ji} := p_j \circ \varphi_{ji}^{-1} \quad \text{for } i \leq j.$$

1. All mappings φ_{ji} being affine and continuous, one has

(1) $\qquad r(p_{ji}) = x_i$ whenever $i \leq j$.

Further

(2) $\qquad m\circ\varphi_{ki}^{-1} = m\circ(\varphi_{ji}\circ\varphi_{kj})^{-1} = (m\circ\varphi_{kj}^{-1})\circ\varphi_{ji}^{-1}$ for $i \leq j \leq k$ and $m \in P(S_k)$.

(3) $\qquad$ The net $(p_{ji})_{j\geq i}$ in $P(S_i)$ decreases in Choquet order.

To see (3) consider $i \leq j \leq k$. Because of (1), we have $r(p_{kj}) = x_j$. S_j being a simplex, proposition 1.5.6 ensures that $p_{kj} < p_j$. Let f be a convex continuous function on S_i; then $f\circ\varphi_{ji}$ is also convex and continuous, hence by (2)

$$\int_{S_i} f\, dp_{ki} = \int_{S_j} f\circ\varphi_{ji}\, dp_{kj} \leq \int_{S_j} f\circ\varphi_{ji}\, dp_j = \int_{S_i} f\, dp_{ji}.$$

2. By theorem 2.4.2, the net $(p_{ji})_{j\geq i}$ converges to its greatest lower bound q_i. In particular, for every $i\in I$ the barycenter of q_i is x_i. Moreover, the net $(q_i)_{i\in I}$ together with the mappings φ_{ji} is a projective system of measures (in the sense of Schwartz(1973)): choose $i\leq j$; then

$$p_{ki} \overset{w}{\to} q_i \quad \text{for} \quad k\geq i \quad \text{and} \quad p_{kj} \overset{w}{\to} q_j \quad \text{for} \quad k\geq j,$$

hence

$$p_{ki} \overset{w}{\to} q_j\circ\varphi_{ji}^{-1} \quad \text{for} \quad k\geq j,$$

which shows that

$$q_i = q_j\circ\varphi_{ji}^{-1}.$$

3. Because the net $(I,\leq)$ is countably generated, the projective limit of the system $(q_i)_{i\in I}$ exists, i.e. there is a unique $p\in P(S)$ such that $q_i = p\circ\varphi_i^{-1}$ for every $i\in I$ (Schwartz(1973), corollary on p. 81 and theorem 21 on p. 75).

4. Now we show, that p is the unique boundary measure on S with barycenter x.

Choose $q\in P(S)$ with $r(q) = x$ and fix $i\in I$. Then $r(q\circ\varphi_i^{-1}) = x_i$; furthermore

$$q\circ\varphi_i^{-1} < p\circ\varphi_i^{-1} = q_i.$$

This relation being true for each $i\in I$ is equivalent to $p<q$, as we will see immediately:

Every continuous linear functional f on $E = \prod_{i\in I} E_i$ is of the form

$$f = \sum_{j \in J} l_j \circ pr_j$$

where J is a finite subset of I, the l_j are in E'_j and the pr_j are the canonical projections (Schaefer(1971), thm. IV.4.3). If k is an upper bound for J then

$$f(x) = \left(\sum_{j \in J} l_j \circ \varphi_{kj}\right) \circ \varphi_k(x) \quad \text{for every} \quad x \in S.$$

Since every $g \in S(E)$ is of the form $\max_{i \leq n}(f_i + c_i)$, $f_i \in E'$, $c_i \in \mathbb{R}$, we see from the last formula that $q < p$ if and only if $q \circ \varphi_i^{-1} < p \circ \varphi_i^{-1}$ for every $i \in I$.

Hence p dominates every $q \in P(S)$ which represents x. This shows that p is the unique boundary measure on S with barycenter x. That S is measure convex was already shown in lemma 3.2.2. □

Remarks.

1. The inverse limit can be empty.
2. Corollary 3.1.3.b is an immediate consequence of theorem 3.2.3.

For sake of completeness, we prove the well-known result for compact simplices (cf. section 3.3).

3.2.4 Proposition. Let $(I, \leq)$ be an increasing net and $(E_i)_{i \in I}$ a family of locally convex spaces. Consider an inverse system (S_i, φ_{ji}) of compact Choquet simplices S_i in E_i. Then the inverse limit $S := \varprojlim S_i$ is a compact Choquet simplex.

Proof. We use the notation from the proof of theorem 3.2.3. Recall that a compact Choquet simplex is the same as a compact simplex (remark 1 in section 1.5). By lemma 3.2.2 S is a convex compact subset of E. Also from the form of S noted in this lemma it is clear, that the measures q_i constructed in part 2 are concentrated on $pr_i[S]$. Hence Prokhorov's theorem (Schwartz(1973), thm. 21 on p. 74) applies to the projective system of measures $(q_i)_{i \in I}$ to ensure the existence of the unique projective limit $p \in P(S)$. Repeating part 4 word by word completes the proof. □

Corollary 3.1.3.a might support the belief that proposition 3.2.4 holds even for closed simplices. Amusingly, almost the same example which showed that closedness cannot be dropped in corollary 3.1.3.a shows now that proposition 3.2.4 does not hold for closed simplices in general.

Example. We use the notation from the example in section 3.1. The set N of Lebesgue null-sets is directed upwards w.r.t. inclusion. For each $N \in N$ the set

$$S_N := P([0,1] \setminus N)$$

is a closed simplex in the locally convex space $M([0,1] \setminus N)$. If

$$\iota_{NL} : [0,1] \setminus N \to [0,1] \setminus L, \; L \subset N \in N,$$

are the canonical embeddings, the simplices S_N, $N \in N$, together with the mappings

$$\varphi_{NL} : S_N \to S_L, \; \mu \to \mu \circ \iota_{NL}^{-1}$$

are an inverse system of closed simplices. By the (easy) proposition IX.5.8 in Bourbaki(1969), the simplices S_N and

$$T_N := \{\nu \in P([0,1]) : \nu(N) = 0\}$$

from the above mentioned example are affinely homeomorphic. By the uniqueness of inverse limits (Semadeni(1971), 11.8.8.(B)), also the inverse limits $\varprojlim S_N$ and TC are affinely homeomorhic. This shows that $\varprojlim S_N$ fails to be a simplex, since TC does so.

Under separability assumptions, we get again integral representations by means of measures which are concentrated on the extreme points of the inverse limit.

<u>3.2.5</u> <u>Corollary</u>. Let $(I,\leq)$ be an increasing net and $(E_i)_{i\in I}$ a family of locally convex spaces. Assume further that (S_i,φ_{ji}) is an inverse system of simplices $S_i \subset E_i$ and that there is a countable subnet $J\subset I$ generating I such that for every $j\in J$ the simplex S_j is a Suslin space.

Then $S := \varprojlim S_i$ is a Suslin simplex. In particular, every element $x\in S$ is represented by one and only one measure $p\in P(\text{ex } S)$.

<u>Proof</u>. We may assume that I itself is countable and every S_i is a Suslin space. Countable products and intersections of Suslin spaces are again Suslin spaces (Schwartz(1973), p. 96). Hence from the form of the inverse limit S - noted in lemma 3.2.2 - we conclude that S is Suslin. It is a simplex by theorem 3.2.3. Application of proposition 1.4.2 completes the proof. □

<u>Remark</u>. The inverse limit of a countable inverse system of metrizable compact simplices is clearly a metrizable compact space. By proposition 3.2.4, it is a metrizable compact Choquet simplex.

Let us finally touch the question, whether inverse limits exist in the category of simplices and affine continuous mappings which preserve extreme boundaries. Only a partial result is obtained . Topological notions not defined here are those from Bourbaki(1966).

<u>3.2.6</u> <u>Proposition</u>. Suppose that the assumptions of corollary 3.2.5 hold true. Assume moreover that $\varphi_{ji}(\text{ex } S_j)\subset \text{ex } S_i$ whenever $j\geq i$. Then $S = \varprojlim S_i$ is a Suslin simplex and ex S is homeomorphic to the topological inverse limit of the inverse system $(\text{ex } S_i, \varphi_{ji}|\text{ex } S_j)$ of topological spaces.

The main property used in the proof is (i) in the following lemma. The idea of the proof is borrowed from Davies/Vincent-Smith(1968), where the compact version of proposition 3.2.6 is shown (cf. the remark at the end of this section).

<u>3.2.7</u> <u>Lemma</u>. Suppose that (S_i,φ_{ji}) is an inverse system of measure convex sets which satisfy

(i) each element x_i in S_i is the barycenter of one and only one measure $p_i \in P(\text{ex } S_i)$.

Assume moreover that $\varphi_{ji}(\text{ex } S_j) \subset \text{ex } S_i$ whenever $j \geq i$. Denote by S the inverse limit of the system. Then $(\text{ex } S_i, \varphi_{ji}|\text{ex } S_j)$ is an inverse system of topological spaces with topological inverse limit ex S.

<u>Proof</u>. It is sufficient to show for $x = (x_i) \in S$ that $x \in \text{ex } S$ if and only if $x_i \in \text{ex } S_i$ for each index i. The rest is standard (cf. Bourbaki(1966)).

If always $x_i \in \text{ex } S_i$ then clearly $x \in \text{ex } S$.

To prove the converse, choose $x \in \text{ex } S$. By (i), there is for each x_i a unique representing $p_i \in P(\text{ex } S_i)$. Since φ_{ji} preserves extreme points, again by uniqueness

(1) $$p_j \circ \varphi_{ji}^{-1} = p_i \quad \text{whenever } i \leq j.$$

Fix now some index i and choose a Borel set B_i in S_i; define for each $j \geq i$ a measure on $\mathcal{B}(\text{ex } S_j)$ by

$$m_j(A_j) := p_j(A_j \cap \varphi_{ji}^{-1}(B_i)), \quad A_j \in \mathcal{B}(S_j).$$

These measures are consistent w.r.t. the mappings φ_{ji} - in fact, for $j \leq k$ applying (1) , we get

(2) $$\begin{aligned} & m_k \circ \varphi_{kj}^{-1}(A_j) = \\ & = p_k\left(\varphi_{kj}^{-1}(A_j) \cap \varphi_{ki}^{-1}(B_i)\right) = p_k\left(\varphi_{kj}^{-1}(A_j) \cap \varphi_{kj}^{-1}(\varphi_{ji}^{-1}(B_i))\right) = \\ & = p_k\left(\varphi_{kj}^{-1}(A_j \cap \varphi_{ji}^{-1}(B_i))\right) = p_j\left(A_j \cap \varphi_{ji}^{-1}(B_i)\right) = \\ & = m_j(A_j). \end{aligned}$$

Define as well measures $\overline{m}_j$, using $S_i \setminus B_i$ instead of B_i. The measures m_j and $\overline{m}_j$ are tight measures with total mass $p_i(B_i)$ and $p_i(S_i \setminus B_i)$

respectively. We claim that

(3) $p_i(B_i) \in \{0,1\}$.

Otherwise, the measures m_j and $\bar{m}_j$, $j \geq i$, may be normalized to get tight probability measures on the sets $\operatorname{ex} S_j$ with barycenters $y_j \neq x_j$ and $z_j \neq x_j$ respectively. Because of (2) the families $(y_j)_{j\geq i}$ and $(z_j)_{j\geq i}$ may trivially be extended to elements $S \ni y \neq x$ and $S \ni z \neq x$; by construction x is a nontrivial convex combination

$$x = p(B_i)y + (1-p(B_i))z,$$

in contradiction to the extremality of x.

By (3), the measure p_i is concentrated on a singleton - so $p_i = \varepsilon_{x_i}$, and from $p_i(\operatorname{ex} S_i)=1$, we conclude that $x_i \in \operatorname{ex} S_i$. This completes the proof. □

<u>Proof of proposition 3.2.6</u>.

The first part is simply corollary 3.2.5. For the rest combine proposition 1.4.2.c and lemma 3.2.7. □

<u>Remark</u>. As already mentioned, proposition 3.2.6 is proved for compact simplices (and arbitrary nets) in Davies/Vincent-Smith(1968), theorem 14. These authors use the following result (loc.cit. lemma 6):

<u>Lemma</u>. If φ is a continuous affine map of the compact simplex S into the compact simplex T, then $\varphi(\operatorname{ex} S) \subset \operatorname{ex} T$ if and only if φ maps maximal measures to maximal measures.

It would be interesting to know to which extent it holds for non-compact simplices.

Section 3.3. The finite dimensional case: a historical note

A. N. Kolmogorov claimed - among others - that the intersection of a decreasing sequence of finite dimensional simplices is itself a finite dimensional simplex. V. Borovikov showed in 1952 that this conjecture is true. Despite the fact that his result follows from more general ones published later (see below), his proof deserves special interest: it is elementary and simple; it can be understood by anyone who has the prerequisites to understand the following definition of a d-simplex. Since Borovikov's paper is only available in Russian, his proof is reported here.

A finite dimensional simplex is a subset of some linear space which is the convex hull of a finite set of affinely independent vectors - its vertices. It is called d-dimensional (or a d-simplex) if it has $d+1$ vertices.

3.3.1 Proposition. The intersection of a decreasing sequence of finite dimensional simplices is a finite dimensional simplex.

Proof (Borovikov(1952)).

Let

$$S_1 \supset S_2 \supset S_3 \supset \dots$$

be the sequence of simplices in question. We may assume that the dimension of all simplices S_i is d and that all the simplices are contained in the Euclidean space $\mathbb{R}^d$. Let $a_i^0,\dots,a_i^d$ be an enumeration of the vertices of S_i. There is a convergent subsequence of $\left((a_i^0,\dots,a_i^d)\right)_{i\in\mathbb{N}}$ with limit $(a^0,\dots,a^d)$; we may and will assume that for every $k = 0,\dots,d$ the sequence $(a_i^k)_{i\in\mathbb{N}}$ converges to some point a^k.

One observes that

$$S := \text{co}\,\{a^k:\ 0 \leq k \leq d\} = \bigcap_{i \in \mathbb{N}} S_i.$$

The inclusion $S \subset \bigcap_{i \in \mathbb{N}} S_i$ is clear. On the other hand, if $\varepsilon > 0$ all but a finite number of the vectors $(a_i^0, \ldots, a_i^d)$ are in

$$U_\varepsilon := \{x \in \mathbb{R}^d : \text{dist}(x,S) < \varepsilon\}$$

and - U_ε being convex - the same is true for the simplices S_i. This proves the converse inclusion.

Let a be an extreme point of S. The vertex a_ℓ^k is said to <u>belong</u> <u>to</u> <u>a</u> if it is a member of some sequence $(a_i^k)_{i \in \mathbb{N}}$ which converges to a. Set

$$C_i^a := \text{co}\left\{a_i^k:\ 0 \leq k \leq d,\ a_i^k \text{ belongs to } a\right\},$$

$$D_i^a := \text{co}\left\{a_i^k:\ 0 \leq k \leq d,\ a_i^k \text{ does not belong to } a\right\}.$$

One has

$$\lim_{i \to \infty} \text{dist}(a, D_i^a) > 0,$$

since this limit is equal to

$$\text{dist}\Big(a, \text{co}\,(\{a^k:\ 0 \leq k \leq d\} \setminus \{a\})\Big) > 0.$$

S has at most $d+1$ extreme points. Let $e^1, \ldots, e^r$, $1 \leq r \leq d+1$, denote the extreme points of S and $u^0, \ldots, u^r$ the unit vectors in $\mathbb{R}^{r+1}$ and T the standard simplex generated by them. There are affine mappings

$$\Theta_i : S_i \to T,\quad i \in \mathbb{N},$$

such that

$$\Theta_i\left[C_i^{e^j}\right] = \{u^j\},\ 1 \leq j \leq r,$$

and vertices of S_i not belonging to any extreme point of S are mapped onto u^o. We show that

$$\Theta_i(e^j) \to u^j,\quad i \to \infty, \text{ for every } 1 \leq j \leq r.$$

Let e be an extreme point of S and write e as a convex combination

$$e = \sum_{k=0}^{d} c_i^k \, a_i^k .$$

Rewriting it in the form

$$e = \sum_{a_i^k \in C_i^e} c_i^k \, a_i^k + \sum_{a_i^k \in D_i^e} c_i^k \, a_i^k ,$$

we see that e is a convex combination

$$e = b_i x_i + (1-b_i) y_i$$

of elements $x_i \in C_i^e$ and $y_i \in D_i^e$. Since

$$\operatorname{dist}(e, C_i^e) \to 0, \quad \operatorname{dist}(e, D_i^e) \to c > 0, \quad i \to \infty,$$

we get

$$b_i \to 1, \; i \to \infty.$$

Assume now that e is the extreme point e^j of S. Then the number b_i is the barycentric coordinate of $\Theta_i(e)$ in T belonging to u^j. Hence $\Theta_i(e^j)$ converges to u^j.

Finally, we conclude that S is a simplex:

Choose $\varepsilon > 0$ such that any displacement of the vertices of T smaller than ε leads to a simplex and $k \in \mathbb{N}$ such that

$$\operatorname{dist}(\Theta_k(e^j), u^j) < \varepsilon, \quad 1 \leq j \leq r.$$

Then $\operatorname{co}\{\Theta_k(e^j) : 1 \leq j \leq r\}$ is an r-simplex, hence also
$S = \operatorname{co}\{e^j : 1 \leq j \leq r\}$. □

Under additional assumptions, the vertices can be labelled in a natural way (cf. Edwards(1975) and the remarks below).

3.3.2 Corollary. Let $(S_i)_{i\in\mathbb{N}}$ be a decreasing sequence of d-simplices in $\mathbb{R}^d$ such that their intersection S has nonvoid interior. Then S is a d-simplex with set of vertices

$$\text{ex } S = \{b^0,\dots,b^d\}$$

and there is an enumeration of the vertices of each S_i

$$\text{ex } S_i = \{b_i^0,\dots,b_i^d\}, \quad i \in \mathbb{N},$$

such that for every $0 \leq j \leq d$ the sequence $(b_i^j)_{i\in\mathbb{N}}$ converges to b^j.

Proof. Take as in the preceding proof a convergent subsequence of $s := \left((a_i^0,\dots,a_i^d)\right)_{i\in\mathbb{N}}$ with limit $(b^0,\dots,b^d)$. The simplex S is the convex hull of $\{b^0,\dots,b^d\}$ and by assumption every b^j is a vertex of S.

Choose $\varepsilon > 0$ and set $U_\varepsilon^j := \{x \in \mathbb{R}^d : \text{dist}(x,b^j) < \varepsilon\}$. Up to a finite number the vertices of all S_i are contained in $\bigcup_{j=0}^{d} U_\varepsilon^j$; otherwise s would have a clusterpoint $(c^0,\dots,c^d)$ different from $(b^0,\dots,b^d)$ and with all c^j vertices of $S = \text{co}\,\{c^0,\dots,c^d\}$, which contradicts $S = \text{co}\,\{b^0,\dots,b^d\}$.

To enumerate the vertices, choose ε such that the neighbourhoods U_ε^j are pairwise disjoint. There is $k \in \mathbb{N}$, such that for every $i \geq k$ exactly one of the vertices of S_i is in U_ε^j. Denote it by b_i^j. □

An inspection of Borovikov's proof shows that the method is confined to finite dimension.

In section 1.5, we noted conditions (Σ), (Σ'), (C) and (L). For compact convex subsets of locally convex spaces they are eqivalent. Hence each of them may be used for the definition of compact Choquet simplices. As soon as one accepts the fact, that a compact set in finite dimension is a d-simplex if and only if it satisfies say condition (L) (cf. the end of section 1.5), one has a short proof of

Borovikov's result.

Proof of proposition 3.3.1 using condition (L).

Assume that the decreasing sequence $(S_i)_{i\in\mathbb{N}}$ is embedded into some positive cone $\mathbb{R}^n_+$ such that its minimal affine extension does not meet zero. Denote its intersection by S. Denote further by C_i and C the cones $\mathbb{R}_+\cdot C_i$ and $\mathbb{R}_+\cdot C$. To show that C is a lattice cone in its own order, it is sufficient to construct for any two elements of C the least upper bound.

Choose x and y in C and consider their least upper bound z_i w.r.t. the own order of C_i. The sequence $(z_i)_{i\in\mathbb{N}}$ increases coordinatewise:

$$z_{i+1} \in \Big((x + C_{i+1}) \cap (y + C_{i+1})\Big) \subset \Big((x + C_i) \cap (y + C_i)\Big) =$$
$$= \Big(z_i + C_i\Big) \subset \Big(z_i + \mathbb{R}^n_+\Big)$$

(the last equality holds because C_i is a lattice cone). Since the sequence is also bounded from above by $x + y$, it converges to some $z \in C$. As

$$C_i \ni (z_i - x) \to (z - x) \in C, \quad i \to \infty,$$

z is an upper bound for x and by the same argument for y in the order of C. For every other upper bound z' we have

$$C_i \ni (z' - z_i) \to (z' - z) \in C.$$

Hence z is even the least upper bound of x and y in the order of C which completes the proof. □

This proof applies to decreasing nets of simplices without change. It is an exercise to give similar proofs by means of the equivalent conditions we mentioned above. These proofs work also for compact Choquet simplices in infinite dimension. The versions with (Σ) and (Σ') are published in Eggleston/Grünbaum/Klee(1964). Davies/Vincent-Smith(1968) used (L) to prove proposition 3.2.4. In a preliminary version of his book on random fields (Preston(1976)), Ch. Preston used their result to describe the structure of certain sets of random fields (personal communication and Edwards(1975), p. 226). We will come back to that aspect in section 4.3. Later on, by a more general approach, Edwards (1974) and (1975) also derived the result of Davies/Vincent-Smith and - apparently not knowing Borovikov's paper - deduced proposition 3.3.1 and corollary 3.3.2. At the same time Kingman(1975a) used the latter results in the examination of anticipation processes. The idea to study stochastic processes by means of their geometric properties is the basis of Kingman(1975b). We will take up this idea in section 4.3.

In all papers mentioned above, the compactness assumption plays an essential role.

CHAPTER 4

INVERSE LIMITS OF SPACES OF MEASURES

The intention of this final chapter is twofold:

- It will be proved that in the category of spaces of measures on standard Borel spaces and substochastic kernels the inverse limits exist. This important measure theoretic result encloses a large class of concrete models arising inevitably in probability theory and in statistics.

- This nontrivial example illustrates the scope of the geometrical results derived in chapter 3. The proof reveals the geometric nature of this - at the first glance purely measure theoretic - situation.

Generalizing the notion of a projective system of measure spaces, we consider inverse systems of spaces of measures. We use the machinery developed in chapter 3 to show that the inverse limits of such systems exist in a setting appropriate for applications. This assertion is formulated in theorem 4.1.3 and plays the central role in the present chapter. It was stated by R. Kotecký and D. Preiss (1979), but a proof was never published. These authors also suggest to consider the abstract theory of Gibbs states in the framework of inverse systems of spaces of measures and sketch how that could be done. This was made more explicit in Kotecký/Preiss(1980), unpublished. In section 4.3 this and other examples from probability theory, statistics and statistical mechanics will be described.

Section 4.1. Inverse limits of spaces of measures and substochastic kernels

Let us start with the precise definition of the objects indicated above.

4.1.1 Definition. Let $(I,\leq)$ be an increasing net. Suppose that for every $i \in I$ a measurable space (X_i, F_i) is given and for $i \leq j$ there is a substochastic kernel P_{ji} from X_j to X_i. Such a family of spaces and kernels is called an inverse system of spaces of measures iff

(i) $P_{ii}(x,\cdot) = \varepsilon_x$ for every $i \in I$ and $x \in X_i$;

(ii) $P_{kj}P_{ji} = P_{ki}$ whenever $i \leq j \leq k$.

A measurable space (X,F) together with substochastic kernels P_i from X to X_i, $i \in I$, is called an inverse limit of this system iff

(iii) $P_i = P_jP_{ji}$ if $i \leq j$;

(iv) if (Y,G) is another measurable space with substochastic kernels R_i from Y to X_i, $i \in I$, such that $R_i = R_jP_{ji}$ whenever $i \leq j$, then there is one and only one substochastic kernel R from Y to X such that $R_i = RP_i$ for every $i \in I$.

Remark. The inverse limits introduced in definition 4.1.1 are unique in the following sense:

if (X,F) and (Y,G) are inverse limits of the same inverse system, then there are stochastic kernels P from X to Y and R from Y to X such that

$$(*) \qquad \begin{aligned} (PR)(x,\cdot) &= \varepsilon_x \quad \text{for every } x \in X, \\ (RP)(y,\cdot) &= \varepsilon_y \quad \text{for every } y \in Y \end{aligned}$$

(Semadeni(1971), 11.8.8(B)).

For the measurable spaces this means measure theoretic isomorphism. A bijection $\Psi : \mathcal{F} \to \mathcal{G}$ between σ-algebras which preserves countable set operations is called a <u>σ-isomorphism</u>. If there is such a map then $\mathcal{F}$ and $\mathcal{G}$ are called <u>σ-isomorphic</u>.

<u>4.1.2</u> <u>Proposition</u>. Suppose that an inverse system of spaces of measures has an inverse limit $(X,\mathcal{F})$ with kernels P_i and an inverse limit $(Y,\mathcal{G})$ with kernels R_i. Then $\mathcal{F}$ and $\mathcal{G}$ are σ-isomorphic.
Assume in addition that $\mathcal{F}$ and $\mathcal{G}$ contain all singletons. Then there is a Borel isomorphism $\psi : X \to Y$ such that

$$P_i(x,\cdot) = R_i(\psi(x),\cdot) \quad \text{for every } x \in X \text{ and } i \in I,$$
$$R_i(y,\cdot) = P_i(\psi^{-1}(y),\cdot) \quad \text{for every } y \in Y \text{ and } i \in I.$$

<u>Proof</u>.

1. The kernels P and R induce mappings

$$\mathcal{B}(Y,\mathcal{G}) \ni g \to P(\cdot,g) \in \mathcal{B}(X,\mathcal{F}),$$
$$\mathcal{B}(X,\mathcal{F}) \ni f \to R(\cdot,f) \in \mathcal{B}(Y,\mathcal{G}),$$

which are monotone - so for instance $g_1 \leq g_2$ implies $P(\cdot,g_1) \leq P(\cdot,g_2)$ - and fulfill in analogy to (*)

$$(**) \qquad \begin{aligned} (PR)(\cdot,f) &= f \quad \text{for } f \in \mathcal{B}(X,\mathcal{F}), \\ (RP)(\cdot,g) &= g \quad \text{for } g \in \mathcal{B}(Y,\mathcal{G}). \end{aligned}$$

We show first that R maps any characteristic function of a measurable set to a characteristic function of a measurable set (so the same is true for P). By monotonicity, we have for $F \in \mathcal{F}$ that $R(\cdot,1_F) \leq 1$ and again by monotonicity and taking notice of (**)

$$1 \geq P(\cdot,1_{\{y:R(y,F)>0\}}) \geq P(\cdot,R(\cdot,1_F)) = 1_F.$$

If $x \notin F$ then (*) implies

$$P(x,\{y: R(y,F) > 0\}) = 0,$$

so that

$$P(\cdot,\{y: R(y,F) > 0\}) = 1_F,$$

hence by (**)

$$R(\cdot,1_F) = 1_{\{y:R(y,F) > 0\}} \in \mathcal{B}(Y,\mathcal{G}).$$

To $F \to R(\cdot,1_F)$ there corresponds therefore canonically a map $\Psi : \mathcal{F} \to \mathcal{G}$ and similarly to $G \to P(\cdot,1_G)$ a map $\Phi : \mathcal{G} \to \mathcal{F}$. From (**) follows

$$\Psi\circ\Phi = id_{\mathcal{G}} \quad \text{and} \quad \Phi\circ\Psi = id_{\mathcal{F}},$$

hence $\Phi = \Psi^{-1}$. Monotonicity of R implies monotonicity of Ψ, i.e. $F_1, F_2 \in \mathcal{F}$, $F_1 \subset F_2$ implies $\Psi(F_1) \subset \Psi(F_2)$. Hence Ψ maps $\mathcal{F}$ onto $\mathcal{G}$ in a one-to-one manner and preserves countable set operations and also its inverse Φ.

2. Assume now that all singletons are measurable. We construct the map ψ directly. Fix $x \in X$. From the first equality in (*), it follows that

$$1 = \varepsilon_x(\{x\}) = \int R(y,\{x\})\, dP(x,y).$$

Hence the measurable set

$$G_x := \{y \in Y: R(y,\{x\}) = 1\} = \{y \in Y: R(y,\cdot) = \varepsilon_x\}$$

has measure one w.r.t. $P(x,\cdot)$. Writing out the second equality in (*) for $y \in G_x$ results in

$$1 = \varepsilon_y(\{y\}) = \int P(x',\{y\})\, d\,R(y,x') = \int P(x',\{y\})\, d\varepsilon_x(x') = \\ = P(x,\{y\}),$$

from which we conclude that $P(x,\cdot) = \varepsilon_y$; in particular G_x is a singleton $\{\psi(x)\}$.

The mapping $\psi : X \to Y$ is well-defined and clearly measurable. In the same way, we construct a measurable mapping $\varphi : Y \to X$ which fulfills

$\varphi\circ\psi(x) = x$ for every $x \in X$ because of the first equality in (*) and by the second equality $\psi\circ\varphi(y) = y$ for every $y \in Y$; thus φ is the inverse of ψ. The rest is clear. □

We state the main result of this chapter. Recall before that a standard Borel space can be defined exclusively in measure theoretic terms (4.3.4).

4.1.3 Theorem. Let $(I,\leq)$ be a countably generated net and let for every $i \in I$ a standard Borel space (X_i, F_i) be given. Assume further that the family $((X_i, F_i))_{i\in I}$ together with substochastic kernels P_{ji} from X_j to X_i, $i \leq j$, is an inverse system of spaces of measures. Then an inverse limit (X,F) together with substochastic kernels P_i, $i \in I$, exists. Moreover, (X,F) can be chosen to be a standard Borel space and if (Y,G) is another inverse limit such that $\{y\} \in G$ for every $y \in Y$ then (X,F) and (Y,G) are Borel isomorphic.

Before we turn to the proof of theorem 4.1.3, we formulate some results of own interest.

4.1.4 Lemma. Suppose that Y is a standard Borel space with σ-algebra G and Z a Polish space with topology r. For every measurable mapping f from Y to Z there is a topology t on Y such that

(i) (Y,t) is a Polish space,

(ii) $G = \mathcal{B}(Y,t)$,

(iii) f is continuous w.r.t. t and r.

Proof. By definition, there is a Polish topology s on the standard Borel space Y such that $G = \mathcal{B}(Y,s)$. On $Y \times Z$ the product of the Borel σ-algebras coincides with the Borel σ-algebra generated by the product topology $s \times r$. Hence by lemma 12 in Schwartz (1973), p. 106, the graph

$$G(f) := \{(y,f(y)) \in Y \times Z:\ y \in Y\}$$

is measurable w.r.t. $\mathcal{B}(Y \times Z, s \times r)$. As a Borel set in the Polish space

$(Y \times Z, s \times r)$, the graph is a Lusin space in the relative topology (loc. cit., thm. 2 on p. 95). So there is a Polish topology t' on $G(f)$ stronger than $G(f) \cap (s \times r)$ and generating the same σ-algebra (loc. cit., cor. 2 on p. 101). Denote by $pr_Y : G(f) \to Y$ and $pr_Z : G(f) \to Z$ the projections. Since pr_Y is a bijection, we can define a Polish topology t on Y by

$$t := \{pr_Y[G]: G \in t'\}.$$

Obviously, we have

$$\begin{aligned} &\mathcal{B}(Y,t) = \\ &= \{pr_Y[B]: B \in \mathcal{B}(G(f),t')\} = \\ &= \{pr_Y[B]: B \in \mathcal{B}(G(f),G(f) \cap (s \times r))\} = \\ &= \mathcal{B}(Y,s) = \\ &= \mathcal{G}. \end{aligned}$$

Writing f in the form $f = pr_Z \circ pr_Y^{-1}$, we see that f as the composition of the t-t'- continuous map pr_Y^{-1} and the t'-r-continuous map pr_Z is continuous w.r.t. the topologies t and r. □

Remark. The assertion can be derived for every topological space Z with countable base and a countable family of maps f from Dellacherie(1980), theorem 40 on page 210.

Next, we note three simple statements about measures on Polish spaces. By Σ we denote the evaluation σ-algebra on $M_+(X,F)$ generated by the maps

$$M_+(X,F) \ni \mu \to \mu(f), \quad f \in \mathcal{B}(X,F).$$

4.1.5 Lemma. Let (X,F) be a Polish space. Then:

a. $M(X,\mathcal{B}(X,t)) = M(X,t)$;

b. $M_+(X,t)$ in its weak topology is a Polish space;

c. Σ coincides with the Borel σ-algebra for the weak topology on $M_+(X,t)$.

Proof.

a. Schwartz(1973), theorem 9 on page 122.

b. Bourbaki(1969), proposition 5.10.

c. The fact that Σ is contained in the Borel σ-algebra follows immediately from Hoffmann-Jørgensen(1970), IV.4.A.7 and the converse holds since the weak topology has a countable base,(Parthasarathy(1967), thm. II.6.6). □

Let V be a pointed convex cone in some locally conves space. A nonempty subset H of V is called a cap iff it is compact and convex and $V \setminus H$ is convex. V is called well-capped iff it is the union of its caps.

4.1.6 Lemma. If (X,t) is completely regular then $M_+(X,t)$ is well-capped.

Proof.

1. For locally compact σ-compact spaces (X,t) the proof can be found in Phelps(1966), proposition 11.5.

2. The general result follows straightforward. Choose $\mu \in M_+(X,t)$ and let $C \subset X$ be the union of pairwise disjoint compact sets C_i, $i \in \mathbb{N}$, such that $\mu(X \setminus C) = 0$. C together with the sum topology t' of the topologies $C_i \cap t$, $i \in \mathbb{N}$, satisfies the conditions in part 1 and has the same Borel sets and tight measures as $C \cap t$ by lemma 2.1.1. Hence, the cone $M_+(C,t')$ is well-capped and $M_+(C,C \cap t)$ as well. Any cap in this cone containing $\mu | \mathcal{B}(C,C \cap t)$ defines a compact convex set in $M_+(X,t)$ containing μ since $M_+(C,C \cap t)$ is affinely homeomorphic to $\{\nu \in M_+(X,t) : \nu(X \setminus C) = 0\}$ (Bourbaki(1969), 5.8). Obviously its complement is convex. □

Remark. Lemma 4.1.6 holds also if (X,t) is an arbitrary separated space (the definition of the weak topology has then to be modified). In v. Weizsäcker/Winkler(1979) a class of subcones of spaces of measures is considered whose members turn out to be well-capped.

The following lemma is due to H. v. Weizsäcker ((1982), unpublished). A ray ρ in a convex cone V is a subset of the form $\{\alpha x: \alpha \geq 0\}$ where $x \in V \setminus \{0\}$. It is called extreme ray iff $x \in \rho$, $y, z \in V$ and $x = y + z$ imply that y and z are proportional. The union of all extreme rays of V is denoted by exr V.

4.1.7 Lemma. If V is a well-capped cone in a locally convex space and V is a Lusin space in the relative topology then exr V is a Lusin space in the relative topology.

Proof. We show that $V \setminus \text{exr } V$ is a Borel set; then exr V is a Lusin space (Schwartz(1973), theorem 2 on page 91).

1. Fix $v \in V$ and choose a cap H containing v. Then $L := \bigcup_{n \in \mathbb{N}} nH$ is σ-compact and all elements $x, y \in V$ with $x + y \in L$ are contained in L (Phelps(1966), page 88; the assumption there that the cone V in question is closed plays no role in the simple proof). Hence

$$R := \{(x,y) \in V \times V: x + y = v\}$$

is a closed subset of $L \times L$ and thus σ-compact.

The set

$$\begin{aligned} &\text{Prop} := \\ &= \{(x,y) \in V \times V: x \text{ and } y \text{ are proportional}\} = \\ &= \bigcup_{\alpha \in [0,1]} \{(x,y) \in V \times V: x = \alpha y \text{ or } y = \alpha x\} \end{aligned}$$

is a closed subset of $V \times V$ (consider convergent nets!).

Hence

$$R_{np} := \{(x,y) \in V \times V: x + y = v, \ x \text{ and } y \text{ are not proportional}\}$$

is a subset of the σ-compact set R and as $R_{np} = R \setminus \text{Prop}$ it is open in R. In a Lusin space, each Borel - in particular each compact - subset is a Lusin space (Schwartz(1973), theorem 2 on page 95) hence a Suslin space and each compact Suslin space is metrizable (loc. cit., cor. 2 on page 106); hence every compact subset of a Lusin space is metrizable. Furthermore, in a metrizable space every open subset is of type F_σ. Putting things together, we get that R_{np} is σ-compact.

2. The set

$$M := \{(x,y,v) \in V \times V \times V: x+y = v,\ x \text{ and } y \text{ are not proportional}\}$$

is a Borel set in $(V \times V) \times V$. For every fixed $v \in V$ the section R_{np} of M is σ-compact. A result of Arsenin (Saint-Raymond(1976), cor.11) then tells us that its projection to the v-axis is a Borel set and thus a Lusin space. □

We are now ready to prove theorem 4.1.3. A technical remark is in order here. For theorem 4.1.3, we will need the results from chapter 3 only for locally convex spaces with the weak topology. Hence also theorem 2.4.2 is only needed in the version for weak topologies - for which we gave a short and simple proof in 2.4.4.

Proof of theorem 4.1.3.

We may and will assume that $I = \mathbb{N}$ and $\leq$ is the natural order since in a countably generated net there is a generating sequence.

1. Choose a Polish topology t_1 on X_1 such that $F_1 = \mathcal{B}(X_1,t_1)$. Recalling lemma 4.1.5, we observe: the space $M_+(X_1,F_1) = M_+(X_1,t_1)$ is Polish in the weak topology; the kernel $P_{2,1}$, being measurable w.r.t. F_2 and the evaluation σ-algebra on $M_+(X_1,F_1)$, is also measurable w.r.t. F_2 and $\mathcal{B}(M_+(X_1,t_1))$. By lemma 4.1.4, there is a Polish topology t_2 on X_2 with $F_2 = \mathcal{B}(X_2,t_2)$ such that $P_{2,1} : X_2 \to M_+(X_1,t_1)$ is continuous. Repeating this procedure, we get a sequence of Polish topologies t_i, $i \in \mathbb{N}$, such that

(i) $F_i = \mathcal{B}(X_i,t_i)$, $i \in \mathbb{N}$,

(ii) $M(X_i,F_i) = M(X_i,t_i)$, $i \in \mathbb{N}$,

(iii) $P_{ji} : X_j \to M_+(X_i,t_i)$, $i \leq j$, is continuous.

2. The sets

$$S_i := \{\nu \in M_+(X_i,F_i) : \nu(X_i) \leq 1\}, \; i \in \mathbb{N} ,$$

are Polish simplices in the locally convex spaces $E_i := M(X_i,t_i)$. Define affine mappings

$$\varphi_{ji} : S_j \to S_i, \; \nu_j \to \nu_j P_{ji}, \; i \leq j.$$

It follows from (iii) that the maps φ_{ji} are continuous. The simplices S_i together with the affine continuous maps φ_{ji} are obviously an inverse system of simplices.

3. Denote by E the product space $\prod_{i\in\mathbb{N}} E_i$ and by τ the product of the topologies $\sigma(M(X_i,t_i),C_b(X_i,t_i))$ on E. Because of corollary 3.2.5, the inverse limit $S := \varprojlim S_i \subset E$ is a Suslin simplex - in fact it is even Polish.

By lemma 4.1.6, each of the cones $M_+(X_i,t_i)$ is well-capped. A straightforward argument shows that the closed subcone $\mathbb{R}_+ \cdot S$ of E is also well-capped (Phelps(1966), proposition 11.4). Since $\mathbb{R}_+ \cdot S$ is a Polish space, the set $\operatorname{exr} \mathbb{R}_+ \cdot S$ is a Lusin space by lemma 4.1.7. From this, we infer that $\operatorname{ex} S \setminus \{0\}$ is a Lusin space. To this end, it is sufficient to observe that

$$\text{(1)} \qquad \operatorname{ex} S = \left(\operatorname{exr} \mathbb{R}_+ \cdot S \cap \{(\nu_i) \in S: \nu_i(X_i) \uparrow 1\}\right) \cup \{0\}.$$

4. Now a candidate for the inverse limit can be defined. Set

$$X := \operatorname{ex} S \setminus \{0\},$$

$$t := X \cap \tau,$$

$$F := \mathcal{B}(X,t) ,$$

$$P_i : X \to M_+(X_i,t_i) = M_+(X_i,F_i), \; x = (\nu_j) \to \nu_i, \; i \in \mathbb{N} .$$

Since $\operatorname{ex} S \setminus \{0\}$ is a Lusin space by part 3, (X,F) is a standard Borel space. The mappings P_i are continuous as projections - hence by lemma 4.1.5 measurable w.r.t. F and the evaluation σ-algebras on the spaces

$M_+(X_i,F_i)$. This shows that they are substochastic kernels from (X,F) to the measurable spaces (X_i,F_i).

5. Set

$$M_{\leq 1} := \{p \in M_+(\text{ex } S \setminus \{0\}) : p(\text{ex } S \setminus \{0\}) \leq 1\}.$$

The barycenter map on $P(\text{ex } S \setminus \{0\})$ is naturally extended to the map

$$\tilde{r} : M_{\leq 1} \to S,\ p \to \tilde{r}(p),$$

where $\tilde{r}(p)$ together with p fulfills the barycentrical formula. Since S is a simplex and because of (1), the map $\tilde{r}$ is an affine bijection and like the barycenter map r continuous. So the map

$$S \ni x \to p_x := \tilde{r}^{-1}(x) \in M_{\leq 1}$$

is the inverse of a continuous bijection between Lusin spaces, hence measurable (Schwartz(1973), lemma 16 on page 107).

6. We shall prove now that (X,F) with kernels P_i defined in part 4 is the inverse limit. By construction

$$P_j P_{ji} = P_i \quad \text{whenever} \quad i \leq j.$$

It is left to check condition (iv) of definition 4.1.1. To this end, consider a space (Y,G) and kernels R_i like there. For every $y \in Y$ define

$$x(y) := (R_i(y))_{i \in \mathbb{N}} \in S.$$

From 5. follows that the map

$$R : Y \to M_+(X,F),\ y \to p_{x(y)}$$

defines a substochastic kernel. Moreover, for every $y \in Y$, $i \in \mathbb{N}$ and $B \in F_i$

$$(RP_i)(y,B) =$$

$$= \int_X P_i(x,B)\ dR(y,x) =$$

$$= \int_{ex\ S \setminus \{0\}} \nu_i(B)\ dp_{(R_j(y))}((\nu_j)) =$$

$$= R_i(y,B).$$

Concerning uniqueness, observe: If a kernel R' as required exists, then by measure convexity $\tilde{r}(R'(y,\cdot)) \in S$ and is necessarily equal to $(R_i(y,\cdot))$. Since S is a simplex, R' equals R; so R is the only kernel satisfying the condition 4.4.1(iv).

7. The remaining part of theorem 4.1.3 was proved in proposition 4.1.2.

□

Remarks.

1. Let us note for later purposes: X consists of those extremal elements (ν_i) of S for which $\nu_i(X_i) \uparrow 1$. This follows from (1) in the preceding proof.

2. An assertion in Scheffer(1971), similar to that of theorem 4.1.3 turns out to be incorrect. A counterexample is provided in Winkler(1984b).

Section 4.2. A criterion for nonemptiness of inverse limits

In the generality of chapter 3, there is no hope to find manageable criterions for nonemptiness of the inverse limit (except trivial ones like compactness). Note that in theorem 4.1.3 the limit space $M(X)$ always contains the zero-measure. So nonemptiness of the inverse limit means the existence of a nonzero measure in $M(X)$. D. Preiss derived a criterion which is presented below. It is contained in the unpublished paper Kotecký/Preiss(1980) which was kindly placed at my disposal by the authors. The formulation is somewhat intricate but the proof is simple. Its applicability will be illustrated in the next section, where a general and well-known theorem on existence of Gibbs states will readily be derived from it.

Some notation is needed. Suppose that (X,F) is any measurable space. For $\mu \in M(X,F)$ the norm $|\mu|$ is the total variation of μ. A linear subspace $A(X)$ of $B(X,F)$ is called <u>norm determining</u> iff for every $\mu \in M(X,F)$ the norm is given by

$$|\mu| = \sup\{\mu(f): f \in A(X), \sup\{|f(x)|: x \in X\} \leq 1\}.$$

The topology on $M(X)$ generated by the functionals $\mu \to \mu(f)$, $f \in A(X)$, is denoted by $\sigma(M(X),A(X))$. Suppose that (Y,G) is another measurable space, $A(X)$ and $A(Y)$ are norm determing, $K \subset M(X)$ and $L \subset M(Y)$. Write K^a and L^a for the closures in the topologies $\sigma(M(X),A(X))$ and $\sigma(M(Y),A(Y))$ respectively. Finally, consider a substochastic kernel Q from Y to X. Then the <u>error in continuity</u> is defined as

$e_{K,L}(Q) := \sup\{|\mu - \nu Q|: \mu \in K^a,\ \nu \in L^a$, there is a net (ν_j) in L with $\nu_j \to \nu$ and $\nu_j Q \to \mu$ in $\sigma(M(Y),A(Y))$ and $\sigma(M(X),A(X))$ respectively$\}$.

<u>4.2.1</u> <u>Proposition</u>. Assume that the standard Borel spaces (X_i,F_i), $i \in \mathbb{N}$, with substochastic kernels P_{ji} are an inverse system of spaces of measures and that the standard Borel space (X,F) with kernels P_i is the inverse limit. Suppose further that for each $i \in \mathbb{N}$ a norm determinig space $A(X_i)$ and a set $M_i \subset M_+(X_i)$ is given such that

(i) for each $i \in \mathbb{N}$ the set M_i is nonvoid, $\sigma(M(X_i),A(X_i))$ relatively compact and P_{ji} maps M_j into M_i whenever $i \leq j$;

(ii) $\lim\limits_{i\to\infty}(\limsup\limits_{j\to\infty} e_{M_i,M_j}(P_{ji})) = 0$;

(iii) there is $a > 0$ and $k \in \mathbb{N}$ such that $\mu(X_k) \geq a$ for each $\mu \in M_k$.

Then the measurable space X is nonempty.

<u>Proof</u>. We may pass over to a subsequence without changing the inverse limit. So, assume that

(1) $$e_{M_i,M_j}(P_{ji}) \leq 2^{-i} \quad \text{for all } i \leq j.$$

There is a sequence $(\nu^{(j)})$ in $M := \prod_{i=1}^{\infty} M_i$ such that

$$\nu_i^{(j)} = \nu_j^{(j)} P_{ji} \quad \text{whenever} \quad i \leq j.$$

Choose some clusterpoint (ν_i) of $(\nu^{(j)})$ in M^a, where M^a is the closure of M in $\prod_{i=1}^{\infty} \sigma(M(X_i), A(X_i))$. From (1) follows that

$$(2) \qquad |\nu_i - \nu_j P_{ji}| \leq 2^{-i} \quad \text{whenever} \quad i \leq j.$$

Set

$$\mu_i^{(j)} := \nu_j P_{ji} \quad \text{for} \quad j > i \quad \text{and} \quad \mu_i^{(j)} = \nu_i \quad \text{for} \quad j \leq i.$$

Then

$$|\mu_i^{(j)} - \mu_i^{(j+1)}| = 0 \quad \text{whenever} \quad i \geq j + 1,$$

and for $i \leq j$

$$\begin{aligned} &|\mu_i^{(j)} - \mu_i^{(j+1)}| = \\ &= |P_{ji}(\nu_j - \nu_{j+1} P_{(j+1),j})| \leq \\ &\leq |\nu_j - \nu_{j+1} P_{(j+1),j}| \leq \\ &\leq 2^{-j}, \end{aligned}$$

where the last inequality is (2) and that before follows from contractivity of the kernels. So there exists $\mu_i := \lim_{j \to \infty} \mu_i^{(j)}$ in norm. By norm continuity of the kernels, one has

$$\mu_j P_{ji} = \mu_i \quad \text{whenever} \quad i \leq j.$$

So there is $\mu \in M_+(X)$ such that $\mu_i = \mu P_i$ for every i. By (iii) and the contractivity of the kernels P_i, it fulfills $\mu(X) \geq a$; hence μ is not the zero-measure. This completes the proof. □

In the proposition, relative compactness of certain sets of measures is demanded. A simple criterion is the following, which proves to be appropriate in the situation of Gibbs states studied in section 4.3.2.

4.2.2 Lemma. Let (X,F) be a measurable space, M a bounded subset of $M_+(X)$ and suppose that for each $b > 0$ there is $\rho_b \in M_+(X)$ and $a > 0$ such that $\rho_b(B) < a$ implies $\nu(B) < b$ whenever $B \in F$ and $\nu \in M$. Then M is relatively compact in the topology $\sigma(M(X), B(X))$.

Proof. Straightforward. □

Section 4.3. Examples from probability theory, statistical mechanics and statistics

A few concrete examples from different fields of stochastics are considered, where inverse systems of spaces of measures arise in a natural way.

The first part deals with entrance boundaries of Markov processes. It may be viewed as an introduction to the more general second example. There, the structure of the set of Gibbs fields for a given specification is studied in some detail. Two important and well-known theorems turn out to be simple consequences of theorem 4.1.3 and proposition 4.2.1. This examples includes theorems like de Finetti's. Then projective statistical fields are sketched and the section is concluded by remarks concerning further models and an open problem.

4.3.1 Entrance boundaries of Markov processes

Let I be a subset of the real line with usual order and $((X_i, F_i))_{i \in I}$ a family of standard Borel spaces. Then every Markov process with time parameter set I and state spaces X_i can be described as a family of stochastic kernels P_{ij}, $i \leq j$ from X_i to X_j satisfying the "Chapman-Kolmogorov" equations

(CK) $$P_{ik} = P_{ij} P_{jk},$$

(see e.g. Kuznecov (1981)). The kernels P_{ij} are called transition

<u>probabilities</u> and $P_{ij}(x,F)$ is interpreted as the probability to hit F at time j having started in x at time i. By (CK), each Markov process can be viewed as an inverse system of spaces of measures (if one does so, one has to take the reversed natural order on I in definition 4.1.1). If (X,F) together with the kernels P_i is the inverse limit, it is called the <u>entrance boundary</u> of the process. An element x in X is interpreted as the beginning of some trajectory of the process or as the entrance through which a path gets into $\prod_{i\in I} X_i$. A measure $\mu \in P(X)$ is the "initial distribution" of the process. If I has a minimum a then the entrance boundary is simply X_a and the probability measures on X_a are the usual initial distributions.

Theorem 4.1.3 shows that for Markov processes with standard Borel state spaces the entrance boundary exists and is again a standard Borel space. From the proof and the remark thereafter, we learn that it consists of the extreme coherent families of probability measures, i.e.

$$X = \text{ex}\ \{(\mu_i)_{i\in I} \in \prod_{i\in I} P(X_i) : \mu_j = \mu_i P_{ij},\ i \leq j\}.$$

Coherent families of probability measures hence are called <u>entrance laws</u>.

This entrance boundary was constructed by E.B. Dynkin (1971). A comprehension of some of its aspects can be found in Dynkin(1978). Under additional assumptions the dual Markov chain - i.e. the process reversed in time - can be used to construct the "exit boundary" of the original process from the entrance boundary of the reversed process. A reference for that is also Dynkin(1978); S.E. Kuznecov (1974) shows where this procedure does not work.

Closely connected is also the following problem. Fix a family $P = (P_{ij})$ of transition probabilities satisfying (CK). One is interested in the structure of the set M(P) of all Markov processes with transition function P, i.e. all those probability measures $\mathbb{P}$ on $X = \prod_{i \in I} X_i$ with

$$\mathbb{P}(A \mid F_{\leq i}) = \mathbb{P}_{i,\xi_i}(A) \quad \text{for every } i \in I,\ A \in F_{>i},\ \mathbb{P}\text{-a.s.}$$

where ξ_i are the projections (i.e. the process), $F_{\leq i}$ is the σ-algebra of the past, generated by all ξ_j, $j \leq i$, $F_{>i}$ is the σ-algebra of the (strict) future, generated by all ξ_j, $j > i$, and $\mathbb{P}_{i,x}$ is the probability measure on $F_{>i}$ with finite dimensional distributuions given by

$$P_{ii_1}(x,dx_1)P_{i_1 i_2}(x_1,dx_2)\ldots P_{i_{n-1}i_n}(x_{n-1},dx_n),$$
$$i_1 < i_2 < \ldots < i_n \in (i,\infty) \cap I.$$

It is easy to see that M(P) is isomorphic to the set of entrance laws in an appropriate sense so that M(P) is a convex set admitting unique integral representations over its extreme boundary.

4.3.2 Specifications and their Gibbs states

The task of statistical mechanics is to derive macroscopic properties of matter from the laws governing the behaviour of the individual particles like atoms or molecules. The methods developed there turned out to be useful also in biology and many other disciplines.

In a first step the equilibrium behaviour of infinite particle systems is studied. Forgetting a lot of structure inherent in particular systems, H. Föllmer (1975) suggested the model described below. It is not only appropriate in the mathematical theory of statistical mechanics. It proved to be fruitful also in probability theory itself and contains and generalizes objects like Markov processes, allows to consider stochastic processes with more dimensional parameter sets; even de Finetti's and related theorems can be studied in this framework. Background material can be found in the introductory chapters of Preston (1976).

Let us first describe Föllmer's model and then introduce a slightly more general one (there are good reasons for the generalization).

Let $(I,\leq)$ be an increasing net which is countably generated. Assume that (X,F) is a standard Borel space and for every $i \in I$ there is a sub-σ-algebra F_i of F such that

$$F_j \subset F_i \quad \text{whenever} \quad i \leq j.$$

A collection Q of stochastic kernels Q_i, $i \in I$, from X to X is called a specification iff

(i) $Q_i(\cdot,F)$ is F_i-measurable for every $F \in F$,

(ii) $Q_i(\cdot,F) = 1_F$ for every $F \in F_i$, $i \in I$,

(iii) $Q_j Q_i = Q_j$ whenever $i \leq j$.

A probability measure μ on F is called a Gibbs state for Q iff it satisfies the DLR - equations

(iv) $\mu Q_i = \mu$ for every $i \in I$.

The last definition originates from Dobrushin, Lanford and Ruelle.

To obtain a rough intuition, think about a lattice model. The set $\mathbb{Z}^d$ might represent the set of locations or sites of physical objects (say iron atoms in the so-called Ising model) which can attain certain states in sets Y_v, $v \in \mathbb{Z}^d$, (say "spin up" or "spin down"). A state of the whole system is then represented by some $x \in X := \prod_{v \in \mathbb{Z}^d} Y_v$. The index set I consists of all bounded subsets of $\mathbb{Z}^d$ and $i \leq j$ iff $i \subset j$. Then F is the product σ-algebra, generated by all projections $\xi_v : X \to Y_v$, F_i is the σ-algebra of events outside i, generated by all ξ_v, $v \notin i$; also the σ-algebra B_i of events inside i, generated by all ξ_v, $v \in i$, is introduced. Given $x \in X$ and an event $F \in B_i$, $Q_i(x, F \times \prod_{v \notin i} Y_v)$ depends only on the coordinates of x outside i because of condition (i) and is interpreted as the probability of the event F inside the bounded vessel i given the boundary condition $(\xi_v : v \notin i)$ outside. The explicit form of Q_i is given by a formula involving the Hamiltonian of the system.

The frame above does not contain certain continuous models where the space of locations is no more discrete, e.g. $\mathbb{R}^d$. The definition of the stochastic kernels Q_i may break down for exceptional sets $N_i \in F_i$ - for $x \in N_i$ one is forced to set $Q_i(x, \cdot) = 0 \in M(X,F)$. This can e.g. happen if x represents a configuration in phase space X such that the number of particles in an expanding volume raises too quickly. Such a modification of Föllmer's model is introduced in Preston(1976), page 16 f.

In the formalism presented now, the σ-algebras F_i and henceforth conditions (i) and (ii) are not essential. Further - since we work with substochastic kernels - $Q_i(x,\cdot)$ may be the zero-measure without more ado. So, let us give the following definition which includes that of Preston just mentioned.

Suppose that $(I,\leq)$ is a countably generated increasing net and (X,F) is a standard Borel space. A collection Q of substochastic kernels Q_i, $i \in I$, from X to X is called a <u>specification</u> iff

(S) $\qquad Q_j Q_i = Q_j$ whenever $i \leq j$,

and $\mu \in P(X,F)$ is called a <u>Gibbs state</u> for Q iff it satisfies the <u>DLR-equations</u>

(DLR) $\qquad \mu Q_i = \mu$ for every $i \in I$.

The set of Gibbs states for a fixed specification will be denoted by G.

To a specification there corresponds an inverse system of measures in a natural way. Set

$$X_i := X, \; F_i := F \quad \text{for every} \quad i \in I.$$
$$P_{ji} := Q_j \quad \text{if} \quad i \leq j.$$

It is formally not quite correct to speak of an inverse system, since $P_{ii}(x,\cdot) = \varepsilon_x$ is not fulfilled. This can easily be overcome, embedding the system in a slightly larger one with this property. It was the idea of Kotecký and Preiss to associate such a system to a specification and then to proceed as we will do now. The following theorem will easily be deduced from theorem 4.1.3:

Theorem. Suppose that (X,F) is a standard Borel space and $\mathcal{Q} = (Q_i)_{i\in I}$ is a specification. Then the following holds.

a. There is a standard Borel space (X_∞, F_∞) and a substochastic kernel Q_∞ from X_∞ to X such that that the map $\mu \to \mu Q_\infty$ is an affine bijection from $P(X_\infty, F_\infty)$ onto G; in particular

$$G = \{\mu Q_\infty : \mu \in P(X_\infty, F_\infty)\},$$
$$G \text{ is a convex set,}$$
$$\text{ex } G = \{Q_\infty(x,\cdot) : x \in X_\infty\}.$$

b. The extreme boundary of G is measurable w.r.t. the evaluation σ-algebra Σ.

c. For every Gibbs state μ there is a unique $p_\mu \in P(\text{ex } G, \text{ex } G \cap \Sigma)$ such that

$$\mu(B) = \int_{\text{ex } G} \nu(B) \, dp_\mu(\nu) \quad \text{for every} \quad B \in F.$$

Part a. of this theorem is proposition 2.4 in Preston(1976), page 25 and thus an analogue of theorem 2.2 there; part b. and c. include theorem 2.1, page 24 loc.cit.. They are proved there by a method due to H. Föllmer (1975) which imitates E.B. Dynkin's construction of the entrance boundary considered in the preceding example. Note that in our proof no use of any Choquet type integral representation theorem is made except the trivial representation of probability measures over point masses.

One can interpret $(X_\infty, \mathcal{F}_\infty)$ as a model for spatial infinity and Q_∞ as "thermodynamic limit" which controlles Gibbs states according to their behaviour at infinity.

<u>Proof of the theorem.</u>

Consider the inverse system of spaces $(X_i, \mathcal{F}_i)$ and kernels P_{ji} corresponding to $(X, \mathcal{F})$ and the specification $\mathcal{Q}$. By theorem 4.1.3 the inverse limit $(X_\infty, \mathcal{F}_\infty)$ with kernels P_i exists and may be assumed to be a standard Borel space.

Observe that all kernels P_i are identical: Choose $x \in X$, $i, j \in I$ and $k \geq i, j$. Then

$$P_i(x,\cdot) = P_k P_{ki}(x,\cdot) = P_k Q_k(x,\cdot) = P_k P_{kj}(x,\cdot) = P_j(x,\cdot).$$

Set $Q_\infty := P_i$.

Recall, that $x \in X_\infty$ is a coherent sequence $(\nu_i) \in \Pi\, M_+(X_i, \mathcal{F}_i)$, which implies

$$\nu_i = \nu_k P_{ki} = \nu_k Q_k = \nu_k P_{kj} = \nu_j \quad \text{whenever} \quad i,j,\ k \in I,\ k \geq i,j,$$

hence

$$\nu_i = \nu,\ i \in I, \text{ for some } \nu \in M_+(X,\mathcal{F}) \text{ and } \nu = \nu Q_i \text{ for each } i.$$

The remark at the end of section 4.1 finally ensures that $\nu(X) = 1$, so that $Q_\infty(x,\cdot)$ is a Gibbs state and obiously the same is true for every μQ_∞, $\mu \in P(X_\infty, \mathcal{F}_\infty)$. By construction of X_∞, each Gibbs state is of this form, so we have

$$\mathcal{G} = \{\mu Q_\infty : \mu \in P(X_\infty, \mathcal{F}_\infty)\}$$

and the rest of part a. is clear.

b. Q_∞ is an injective map from the standard Borel space $(X_\infty, \mathcal{F}_\infty)$ into the countably generated space $(P(X), \Sigma)$, so theorem 2.4 in Parthasarathy (1967), page 135 tells us that $\text{ex}\, \mathcal{G} = \{Q_\infty(x,\cdot) : x \in X_\infty\}$ is in Σ.

c. Suppose that $\nu \in G$ and $\nu = \mu Q_\infty$ with $\mu \in P(X_\infty)$. Set $p_\mu := \mu \circ Q_\infty^{-1}$ to get the integral representation. Uniqueness follows from the uniqueness of μ. □

Next, proposition 4.2.1 is applied to the problem of existence of Gibbs states. Theorem 3.3 from Preston(1976) is reformulated and reproved. The result is taken from Kotecký/Preiss(1980).

Some notation has to be introduced. Consider in addition to I a second index set Γ equipped with a partial order $\leq$ which is directed upwards and countably generated. Let $(B_\gamma)_{\gamma \in \Gamma}$ be an increasing net of sub-σ-algebras of F, i.e.

$$B_\beta \subset B_\gamma \quad \text{whenever} \quad \beta \leq \gamma,$$

such that F is the smallest σ-algebra containing all B_γ. Denote by A the algebra $\bigcup_{\gamma \in \Gamma} B_\gamma$, and by A_δ the set of all countable intersections of elements of A. $B(X,A)$ is the the linear space of all real bounded functions on X which are measurable w.r.t. some B_γ.

Usually, in the situation where each $i \in I$ is a bounded subset of $\mathbb{Z}^d$ as sketched above and where F_i consists of events measurable outside i, one has $\Gamma = I$ and B_i is the σ-algebra of events measurable inside i.

<u>Theorem</u>. Suppose that (X,F) is a standard Borel space and Q is a specification. Let $(U_n)_{n \in \mathbb{N}}$ be an increasing sequence of elements from A_δ. Suppose further that for some $x \in X$ the following conditions are satisfied:

(i) for every $\gamma \in \Gamma$ and $b > 0$ there are $\rho \in P(X,B_\gamma)$, $i \in I$ and $a > 0$ such that $\rho(B) < a$, $B \in B_\gamma$ implies $Q_j(x,B) < b$ for all $j \geq i$;

(ii) for every $a > 0$ there is $n \geq 1$ such that $Q_i(x,U_n) \geq 1 - a$ for every $i \in I$;

(iii) whenever $i \in I$, $f \in B(X,A)$, $a > 0$ and $n \geq 1$, there is $g \in B(X,A)$ such that

$$|Q_i(y,f) - g(y)| < a \quad \text{for all} \quad y \in U_n.$$

<u>Proof</u>. Since $(I,\leq)$ is countably generated, we may and will assume that I is the set of natural numbers and $\leq$ the usual order. Proposition 4.2.1 will be applied with $A(X_i) = B(X,A)$ and $M_i = \{Q_j(x,\cdot): j \geq i\}$.

1. Let us first check that condition 4.2.1(i) is satisfied. All is clear except the compactness assumption.

Fix $\gamma \in \Gamma$. Because of condition (i) above, lemma 4.2.2 applies to (X,B_γ) and $\{Q_j(x,\cdot)|B_\gamma: j \geq i\}$ and we infer that each M_i is relatively compact in $\sigma(M(X,F),B(X,B_\gamma))$.

Since the latter holds for every $\gamma \in \Gamma$, the sets M_i are relatively compact in $\sigma(M(X,F),B(X,A))$. In fact, the spaces

$$(M_+(X,B_\gamma),\sigma(M_+(X,B_\gamma),B(X,B_\gamma)))$$

are a projective system of topological spaces together with the restriction maps

$$M_+(X,B_\delta) \ni \nu \to \nu|B_\gamma \in M_+(X,B_\gamma), \quad \gamma \leq \delta.$$

The projective limit is $(M_+(X,F),\sigma(M_+(X,F),B(X,A)))$ together with restrictions. So relative compactness of M_i in the latter space is ensured (Bourbaki(1966), proposition 8 in chapter I, §9.6).

2. Consider now condition 4.2.1(ii). Choose M_i and M_j, $j \geq i$, and a a net $(Q_{i_\theta}(x,\cdot))_\theta$, $i_\theta \geq j$, with $M_j \ni Q_{i_\theta}(x,\cdot) \to \mu \in M_j^a$ in $\sigma(M(X),B(X,A))$. Then

$$Q_{i_\theta}P_{ji}(x,\cdot) = Q_{i_\theta}Q_j(x,\cdot) = Q_{i_\theta}(x,\cdot) \to \mu.$$

So we have to estimate the quantity $|\mu - \mu Q_j|$.

For every $i \in I$ and $a > 0$ there are $f, g \in \mathcal{B}(X, A)$, $n \geq 1$, and θ with $i_\theta \geq i$ such that the following inequalities hold; the single steps will be justified immediately below:

$$
\begin{aligned}
&|\mu - \mu Q_j| \leq \\
&\leq a + |\mu(f) - \mu Q_j(f)| \leq \\
&\leq 2a + 2\mu(X \setminus U_n) + |\mu(f) - \mu(g)| \leq \\
&\leq 4a + 2\mu(X \setminus U_n) + |Q_{i_\theta}(x,f) - Q_{i_\theta}(x,g)| \leq \\
&\leq 5a + 2\mu(X \setminus U_n) + 2Q_j(x, X \setminus U_n) \leq \\
&\leq 7a + 2\mu(X \setminus U_n)
\end{aligned}
$$

where we have used in this order that $\mathcal{B}(X,A)$ is norm determining, (iii), the assumption on μ, the estimate in (iii) and (ii).
Since $U_n \in A_\delta$, for every $c > 0$ there are $A \in A$ and $\tilde{\theta}$ such that

$$
\begin{aligned}
&\mu(U_n) \geq \mu(A) - c \geq Q_{i_{\tilde{\theta}}}(x,A) - 2c \geq Q_{i_{\tilde{\theta}}}(x,U_n) - 2c \geq \\
&\geq 1 - a - 2c;
\end{aligned}
$$

hence $\mu(X \setminus U_n) \leq a$. This together with the above estimate shows that assumption (ii) in proposition 4.2.1 is fulfilled.
Finally, assumption 4.2.1(iii) is contained in (ii) above. Hence the present assertion follows from proposition 4.2.1. □

There are reasonable specifications with no Gibbs states. Such an example was constructed by F.L. Spitzer (cf. Kemeny/Snell/Knapp(1976), example 12-38).

4.3.3 Projective statistical fields

We restrict ourselves to few remarks, since this topic is completely described in chapter IV of S.L. Lauritzen's book (1982) in the same language we used. In this context, an inverse system of spaces X_i and kernels P_{ji} together with a parameter space Θ and kernels μ_i from Θ to X_i such that $\mu_j P_{ji} = \mu_i$ is called a projective statistical field. This is a natural generalization of an "extremal family". The X_i are thought of as sample spaces of statistical experiments subsequently extended in a consistent way - which is expressed by the coherent kernels P_{ji} - and for each experiment one has a usual statistical model: some random variable ξ_i with values in X_i is distributed according to $\mu_i(\theta,\cdot)$ where $\theta \in \Theta$ is unknown.

Such a field is transformed to a new one via a "sufficient" family (t_i) of statistics $t_i : X_i \to Y_i$ (loc. cit. page 200 f.). So one gets a projective statistical field of spaces Y_i and kernels Q_i. If Θ together with kernels ν_i from Θ to Y_i (where $\nu_i(\theta,\cdot) = \mu_i(\theta,\cdot)\circ t_i^{-1}$) is the inverse limit in the sense of definition 4.1.1 then it is called canonical statistical field (loc. cit. page 201). Hence for standard Borel spaces, theorem 4.1.3 assures the existence of such canonical statistical fields. Lauritzen assumes the spaces to be Polish but standard Borel spaces seem to be the natural setting.

In the same spirit repetitive structures - which are introduced and investigated by P. Martin-Löf (1974) - can be studied.

Lauritzen proved a special case of theorem 4.1.3 under topological assumptions - he assumed the spaces to be Polish and the kernels to be continuous. Before, he had derived this and similar results from an assertion in Scheffer(1971) which unfortunately turns out to be incorrect as J. Vestergaard pointed out (personal communication; cf. the remark on page 237 in Lauritzen(1982)). A counterexample is constructed in Winkler(1984b).

4.3.4 Further examples, complements and an open problem

Let us conclude this chapter with three remarks.

Further examples and related topics

Example 4.3.2 by itself encloses a whole class of particular models. Perhaps the most renowned one is de Finetti's theorem. In Georgii(1979) and Accardi/Pistone(1982) the appropriate specification is constructed. There is an immense literature on related topics. It would be carrying things too far to enumerate only part of it. Let us only mention the papers Campanino/Spizzichone(1981) and Accardi/Pistone(1982). An outline is given in Aldous(1984) and a collection of more recent development is Koch/Spizzichone eds.(1982).

We also have to notice that there are different abstract approaches whose scope is also not restricted to the theory of Gibbs states. The main ideas already appear in the parallel but independent papers of R.H. Farrell (1962) and V.S. Varadarajan(1963). Maitra(1977) and Dynkin (1978) follow the same lines (cf. the next subsection). Independently, J. Kerstan (1975) and A. Wakolbinger (1979) worked in the same field which led to the joint papers Kerstan/Wakolbinger(1980), (1981).

Dynkin's B-spaces are standard Borel and vice versa

The most frequently cited of these papers is Dynkin(1978). To clarify the connection with the results presented in this chapter, a remark is in order. Throughout, we assumed the underlying measurable spaces to be standard Borel. Dynkin introduces what he calls B-spaces which - as he remarks - are more general. In fact, the two notions *coincide*. Let us give a proof.

A measurable space (X,F) is a B-space iff there is a countable family H of bounded measurable functions on X - called a support system - such that

(i) if $(\mu_n)_{n\in\mathbb{N}}$ is a sequence in $P(X,F)$ such that

$l(h) := \lim_{n\to\infty} \mu_n(h)$ exists for every $h \in H$ then there is $\mu \in P(X,F)$ with $\mu(h) = l(h)$ for every $h \in H$;

(ii) $B(X,F)$ is the smallest class of functions on X containing H and being closed under addition, scalar multiplication and uniformly bounded convergence.

Proposition. Let (X,F) be a measurable space such that F separates points. Then (X,F) is a B-space if and only if it is a standard Borel space.

Proof.

1. Standard Borel spaces have a support system (Parthasarathy(1967), page 145) so they are B-spaces.

2. Assume now that (X,F) is a B-space. The measurable space $(P(X,F),\Sigma)$ where Σ is the evaluation σ-algebra is a standard Borel space. To see that note first, that Σ is generated by the evaluation functionals $P(X,F) \ni \nu \to \nu(h)$, $h \in H$, because of (ii). Let $H = \{h_1,\dots\}$ so that $\iota : P(X,F) \to \mathbb{R}^{\mathbb{N}}$, $\nu \to (\nu(h_m))_{m\in\mathbb{N}}$ is one-to-one by (ii) and by the above observation a Borel isomorphism of the spaces $(P(X,F),\Sigma)$ and $(\iota(P(X,F)),\iota(P(X,F)) \cap \mathbb{B}(\mathbb{R}^{\mathbb{N}}))$. Because of (i), the space $\iota(P(X,F))$ is closed in $\mathbb{R}^{\mathbb{N}}$, hence standard Borel.

The set X' of Dirac measures on F is measurable: Choose a sequence $F_1, F_2,\dots$ generating F; such a choice is possible because of (ii). Then

$$X' = \bigcap_{n\in\mathbb{N}} \{\mu \in P(X,F): \mu(F_n) \in \{0,1\}\} \in \Sigma.$$

Hence $(X',X' \cap \Sigma)$ is a standard Borel space. This space is Borel isomorphic to (X,F) which implies that (X,F) is a standard Borel space. □

This proof is taken from v. Weizsäcker/Winkler(1980).

An open problem

Let us explain the problem by the example of Gibbs states. Consider a standard Borel space (X,F), a specification $(Q_i)_{i \in I}$ as in example 4.3.2 and its Gibbs states G; at this place we need the σ-algebras F_i introduced there. Define the tail-field

$$F_{\infty-} := \bigcap_{i \in I} F_i.$$

The usual proceeding runs as follows. One constructs a stochastic kernel $Q_{\infty-}$ from $(X,F_{\infty-})$ to (X,F) such that

$$\mu\{x \in X: Q_{\infty-}(x,\cdot) \in G\} = 1 \quad \text{for every} \quad \mu \in G,$$
$$\mathbb{E}^{\mu}(1_F | F_{\infty-}) = Q_{\infty-}(\cdot,F) \quad \text{for every} \quad F \in F \quad \text{and} \quad \mu \in G,$$

where $\mathbb{E}^{\mu}$ is the expectation w.r.t. μ. In other words, $F_{\infty-}$ is a sufficient σ-algebra for G and $Q_{\infty-}$ the corresponding common conditional distribution. Then one shows

$$\text{ex } G \in \{Q_{\infty-}(x,\cdot): x \in X\},$$
$$\mu\{x \in X: Q_{\infty-}(x,\cdot) \in \text{ex } G\} = 1 \quad \text{for every} \quad \mu \in G,$$
$$G = \{\mu \in P(X,F): \mu Q_{\infty-} = \mu\}.$$

From this, an analogue of the first theorem in 4.3.2 is derived (cf. Dynkin(1978), theorem 3.1). As soon as $F_{\infty-}$ and $Q_{\infty-}$ are known, the space (X_{∞},F_{∞}) and the kernel Q_{∞} are easily constructed: Call x and y in X equivalent and write $x \sim y$ iff $Q_{\infty-}(x,\cdot) = Q_{\infty-}(y,\cdot)$. Set $X_{\infty} := X/\sim$; then define F_{∞} and Q_{∞} canonically.

The question is how to go the other way round.

Problem. Construct $Q_{\infty-}$ from Q_{∞}!

Presumably, this boils down to show that the measures $Q_{\infty}(x,\cdot)$ are orthogonal in a strict sense.

Such orthogonality properties are investigated in Mauldin/Preiss/ v. Weizsäcker(1983); but the paper does not provide a solution.

APPENDIX

A disintegration result appropriate for our purposes is proved (theorem A2 below). It is similar to that in Valadier(1973); but the additional measurability property i) is convenient in our context. In the present form it is due to H. v. Weizsäcker ((1977), Satz III.11). An important consequence is the existence of conditional expectations (cf. v. Weizsäcker(1977), Lemma III.27) which is proved here as proposition A3; it was needed for proposition 2.3.5. The disintegration is used in the proof of theorem 1.3.6. It should be noted that related ideas appear in Edgar(1978); see also the references there.

Let us recall some basic notions from the theory of vector measures. The latter are used in A1 and A2.

Suppose that $(X,\mathcal{F})$ is a measurable space, E a locally convex space and m a set function defined on the σ-algebra $\mathcal{F}$ with values in E. We say that m is a <u>vector measure</u> iff $m(\emptyset) = 0 \in E$ and m is weakly countably additive, i.e. $l \circ m$ is countably additive for each $l \in E'$. Suppose further that $\mathbb{P}$ is a probability measure on $\mathcal{F}$. We shall say that m is <u>absolutely continuous w.r.t. $\mathbb{P}$</u> iff $\mathbb{P}(B) = 0$ implies that $m(B) = 0 \in E$ for every measurable set B.

A short digression is in order here. In the remark concluding section 1.2, the Radon-Nikodym Property was mentioned. It is sufficient to formulate it for Banach spaces. Suppose that F is a Banach space. A set function m on $\mathcal{F}$ with values in F and $m(\emptyset) = 0 \in F$ is called <u>strongly countably additive</u> iff it is countably additive in norm; that is whenever $(B_i)_{i \in \mathbb{N}}$ is a sequence of pairwise disjoint elements of $\mathcal{F}$ with union B, then

$m(B) = \sum_{i=1}^{\infty} m(B_i)$ (equivalently the partial sums $\sum_{i \leq n} m(B_i)$ converge in norm to $m(B)$). Absolute continuity is defined as above. A closed bounded convex subset C of F has the <u>Radon-Nikodym Property</u> iff for each probability space $(X, \mathcal{F}, \mathbb{P})$ and every strongly countably additive set function m on $\mathcal{F}$ with values in F which is absolutely continuous w.r.t. $\mathbb{P}$ and has its <u>average range</u>

$$\left\{\frac{m(B)}{\mathbb{P}(B)} : B \in \mathcal{F},\ \mathbb{P}(B) > 0\right\}$$

contained in C there is a function f - Bochner integrable for $\mathbb{P}$ - such that $m(B) = \int_B f \, d\mathbb{P}$ for each $B \in \mathcal{F}$. The space F itself has the Radon-Nikodym Property iff each closed bounded convex subset has the Radon-Nikodym Property.

Let us return to the disintegration.

<u>A1</u>. <u>Lemma</u>. Let $(X, \mathcal{F}, \mathbb{P})$ be a complete probability space and Y a completely regular topological space. Suppose further that m is a vector measure on $\mathcal{F}$ with values in $\mathcal{M}_+(Y)$ absolutely continuous w.r.t. $\mathbb{P}$ and

$$s := \sup\left\{\frac{\|m(F)\|}{\mathbb{P}(F)} : F \in \mathcal{F},\ \mathbb{P}(F) > 0\right\} < \infty,$$

where $\|\cdot\|$ is the dual-space norm on $C_b(Y)'$. Then there is a mapping

$$v : X \to \mathcal{M}_+(Y)$$

with the properties

(i) v is $\mathcal{F}-\mathcal{B}(\mathcal{M}(Y), \sigma(\mathcal{M}(Y), C_b(Y))$ measurable,

(ii) $\mathbb{P} \circ v^{-1} \in \mathcal{P}(\mathcal{M}_+(Y))$,

(iii) v is a density, i.e.

$$\int_F v(x)(B) \, d\mathbb{P}(x) = m(F)(B) \quad \text{for all } B \in \mathcal{B}(Y) \text{ and } F \in \mathcal{F}.$$

<u>Proof</u>. Assume for the moment that Y is compact. Theorem 4.2 in Gold-

man(1977) or results in Kupka(1972) tell us that there is a "weak density", i.e. an $\mathcal{F}-\mathcal{Z}(H)$ measurable map

$$w : X \to H := \{\nu \in \mathcal{M}_+(Y) : \|\nu\| \leq s\},$$

such that

$$(1) \qquad \int_F w(x)(\varphi)\, d\mathbb{P}(x) = m(F)(\varphi) \quad \text{for all } F \in \mathcal{F},\ \varphi \in C_b(Y)$$

(observe that H is compact).
Since H is compact, $\mathcal{Z}(H) = \mathcal{B}_o(H)$ according to the lemma in 0.7 and the lattice version of the Stone-Weierstraß theorem. Thus $\mathbb{P}\circ w^{-1}$ is a Baire measure on H. Apply the extended version of Edgar's weak section theorem due to S. Graf ((1980), corollary 4.8 and remark 4.9.a) on page 130) to $\mathbb{P}\circ w^{-1}$ and the identity on H to see that w may be assumed to be $\mathcal{B}(\mathcal{M}(Y))$-measurable and $\mathbb{P}\circ w^{-1} \in \mathcal{P}(H)$.
For $F \in \mathcal{F}$ with $\mathbb{P}(F) > 0$ define the conditional probability

$$\mathbb{P}_F := \mathbb{P}(F)^{-1}\, \mathbb{P}(F \cap \cdot\,).$$

The image measure $\mathbb{P}_F\circ w^{-1}$ is absolutely continuous w.r.t. $\mathbb{P}\circ w^{-1}$, hence in $\mathcal{P}(H)$. From (1) we read off that $r(\mathbb{P}_F\circ w^{-1})$ is $\mathbb{P}(F)^{-1}m(F)$ and by proposition 1.1.2 for every $B \in \mathcal{B}(Y)$ we have

$$\begin{aligned} &\int_F w(x)(B)\, d\mathbb{P}(x) = \\ &= \mathbb{P}(F) \int_H \nu(B)\, d\mathbb{P}_F\circ w^{-1}(\nu) = \\ &= \mathbb{P}(F)\, r(\mathbb{P}_F\circ w^{-1})(B) = \\ &= m(F)(B); \end{aligned}$$

clearly the outer equality holds for *every* $F \in \mathcal{F}$ which proves the lemma for compact Y.

Let now Y be completely regular. Apply what we have just proved to the Stone-Čech compactification βY and the vector measure $\overline{m}$ defined by

$$\overline{m}(F)(B) := m(F)(B \cap Y),\ F \in \mathcal{F},\ B \in \mathcal{B}(\beta Y)$$

to get a mapping

$$w : X \to M_+(\beta Y),$$

which is Borel measurable, has tight image measure $\mathbb{P} \circ w^{-1}$ and satisfies

(2) $\int_F w(x)(B)\, d\mathbb{P}(x) = m(F)(B \cap Y)$ for each $B \in \mathcal{B}(\beta Y)$ and $F \in \mathcal{F}$.

Since $m(X)$ is tight on Y, there is a σ-compact subset C of Y with $m(X)(Y \setminus C) = 0$. Since C is a Borel set in βY, because of (2)

$$\int_X w(x)(\beta Y \setminus C)\, d\mathbb{P}(x) = m(X)(Y \setminus C) = 0$$

and consequently

$$\mathbb{P}(\{x \in X : w(x)(\beta Y \setminus C) = 0\}) = 1.$$

Set now

$$v(x)(B) := w(x)(B \cap C) \quad \text{for every } x \in X \text{ and } B \in \mathcal{B}(Y)$$

to get a map with the desired properties. □

If X and Y are sets, denote by pr_X and pr_Y the projections from $X \times Y$ onto X and Y respectively.

A2. Theorem. Let $(X, \mathcal{F}, \mathbb{P})$ be a complete probability space and Y a completely regular topological space. Suppose further that μ is a probability measure on the product space $(X \times Y, \mathcal{F} \otimes \mathcal{B}(Y))$ such that $\mu \circ pr_X^{-1} = \mathbb{P}$ and $\mu \circ pr_Y^{-1}$ is tight. Then there is a mapping

$$v : X \to P(Y)$$

with the properties

(i) v is $\mathcal{F} - \mathcal{B}(M(Y), \sigma(M(Y), C_b(Y)))$ - measurable,

(ii) $\mathbb{P} \circ v^{-1} \in P(P(Y))$,

(iii) for each $h : X \times Y \to \mathbb{R}$ in $L^1(\mu)$ there is $N \in F$ with $\mathbb{P}$-measure zero, such that

$h(x,\cdot) \in L^1(v(x))$ for each $x \notin N$,

$X \setminus N \ni x \to \int_X h(x,\cdot)\, dv(x)$ is measurable w.r.t. $(X \setminus N) \cap F$

and

$$\int_{X\times Y} h(x,y)\, d\mu(x,y) = \int_{X\setminus N} \left[\int_Y h(x,y)\, dv(x)(y)\right] d\mathbb{P}(x).$$

Proof. Define a vector measure $m : F \to M_+(Y)$ by

$$m(F)(B) := \mu(F \times B) \quad \text{for every } F \in F \text{ and } B \in \mathcal{B}(Y).$$

Since m is absolutely continuous w.r.t. $\mathbb{P}$ by definition and

$$\mathbb{P}(F)^{-1}\|m(F)\| = \mathbb{P}(F)^{-1}m(F)(Y) = 1 \text{ whenever } \mathbb{P}(F) > 0,$$

m satisfies the hypothesis of the preceding lemma. Condition (iii) there reads now

(1) $$\int_F v(x)(B)\, d\mathbb{P}(x) = \mu(F \times B) \quad \text{for all } F \in F \text{ and } B \in \mathcal{B}(Y);$$

set $B = Y$ to see that

$$\int_F v(x)(Y)\, d\mathbb{P}(x) = \mu(F \times Y) = \mathbb{P}(F),$$

hence v actually maps X into $P(Y)$ almost surely.

In consequence

$$\mathbb{P} \circ v^{-1} \in P(P(Y));$$

this shows that (ii) is true. Further, condition (i) of the lemma and the theorem are identical.

It remains to check the last condition. By (1), it holds for functions of the form $h = 1_{F \times B}$, $F \in F$, $B \in \mathcal{B}(Y)$. Use now the generalized Fubini theorem (Neveu(1969), page 98) to extend it first to functions

$h \in \mathcal{B}(X \times Y, \mathcal{F} \otimes \mathbf{B}(Y))$ and then to functions $h \in L^1(\mu)$. This completes the proof of the theorem. □

The theorem is applied to construct conditional expectations *with tight image measures*.

<u>A3</u>. <u>Proposition</u>. Consider a measure convex subset M of some locally convex space. Suppose that $(X, \mathcal{F}, \mathbb{P})$ is a probability space, $\mathcal{G}$ a $\mathbb{P}$-complete sub-σ-algebra of $\mathcal{F}$ and $f : X \to M$ an $\mathcal{F} - \mathbf{B}(M)$ measurable function with $\mathbb{P} \circ f^{-1} \in \mathcal{P}(M)$. Then there is a $\mathcal{G} - \mathbf{B}(M)$ measurable function $g : X \to M$ such that

$$\left(\mathbb{P} \mid \mathcal{G}\right) \circ g^{-1} \in \mathcal{P}(M) \quad \text{and} \quad g = \mathbb{E}(f \mid \mathcal{G}).$$

<u>Proof</u>. Define the $\mathcal{F} - \mathcal{F} \otimes \mathbf{B}(M)$ measurable map

$$f^* : X \to X \times M, \; x \to (x, f(x)).$$

Let μ denote the restriction of $\mathbb{P} \circ (f^*)^{-1}$ to $\mathcal{G} \otimes \mathbf{B}(M)$. By the theorem, there is a map $v : X \to \mathcal{P}(M)$ with properties (i) - (iii) there. From (iii) one gets for each $G \in \mathcal{G}$ and $\varphi \in C_b(M)$ that

(1) $$\int_G \left[\int_M \varphi(y) \; dv(x)(y)\right] d\mathbb{P}(x) =$$

$$= \int_{X \times M} 1_G(x)\varphi(y) \; d\mu(x,y) =$$

$$= \int_{X \times M} 1_G(x)\varphi(y) \; d\mathbb{P} \circ (f^*)^{-1}(x,y) =$$

$$= \int_G \varphi \circ f(x) \; d\mathbb{P}(x) .$$

Recall that the barycenter map $p \to r(P)$ maps $\mathcal{P}(M)$ into M since M is measure convex, and that r is continuous by proposition 1.1.3. Hence in view of A2.(i) the map $g := r \circ v$ is $\mathcal{G} - \mathbf{B}(M)$ measurable and has tight image measure because of A2.(ii). Finally, for $G \in \mathcal{G}$ and $l \in E'$, the barycentrical formula together with (1) yields

$$\int_G l \circ g(x)\ d\mathbb{P}(x) =$$

$$= \int_G l\Big(r(v(x))\Big)\ d\mathbb{P}(x) =$$

$$= \int_G \left[\int_M l(y)\ dv(x)(y)\right] d\mathbb{P}(x) =$$

$$= \int_G l \circ f(x)\ d\mathbb{P}(x) ,$$

which means that $g = \mathbb{E}(f|G)$. □

REFERENCES

[1] Accardi L. and Pistone G. (1982): *de Finetti's theorem, sufficiency, and Dobrushin's theory*, in: Exchangeability in Probability and Statistics, G. Koch and F. Spizzichino eds., 125-156, North-Holland: Amsterdam

[2] Aldous D.J. (1984): *Exchangeability and related topics*, manuscript, Dpt. of Statistics, University of California, Berkeley

[3] Alfsen E. (1971): *Compact Convex Sets and Boundary Integrals*, Springer Verlag: Berlin, Heidelberg, New York

[4] Asimov L. and Ellis A.J. (1980): *Convexity theory and its applications in functional analysis*, Academic Press: London, New York

[5] Bahadur R.R. and Zabell S.L. (1979): *Large deviations of the sample mean in general vector spaces*, Ann. Probab. 7, 587-621

[6] Bessaga C. and Pelczyński A. (1975): *Selected topics in infinite-dimensional topology*, PWN - Polish Scientific Publishers: Warszawa

[7] Bishop E. and de Leeuw K. (1959): *The representation of linear functionals by measures on sets of extreme points*, Ann. Inst. Fourier (Grenoble) 9, 303-331

[8] Blackwell D. (1951): *Comparison of experiments*, Proc. Second Berceley Symp. Math. Stat. Probab., Univ. of California Press: Berceley, Los Angeles, 93-102

[9] Borovikov V. (1952): *On the intersection of a sequence of simplices*, (Russian), Uspehi Mat. Nauk 7.6, 179-180

[10] Bourbaki N. (1966): *General topology, part 1*, Hermann: Paris

[11] Bourbaki N. (1969): *Eléments de mathématique*, Livre VI, *Intégration*, Chap. IX, Hermann: Paris

[12] Bourgin R.D. (1971): *Barycenters of measures on certain noncompact convex sets*, Trans. Amer. Math. Soc. 154, 323-340

[13] Bourgin R.D. (1979): *Partial orderings for integral representation in convex sets with the Radon-Nikodym Property*, Pacific J. Math. 81, 29-44

[14] Bourgin R.D. (1983): *Geometric aspects of convex sets with the Radon-Nikodym Property*, Lecture Notes in Mathematics 993, Springer Verlag: Berlin, Heidelberg, New York

[15] Bourgin R.D. and Edgar G.A. (1976): *Noncompact Simplexes in Banach Spaces with the Radon-Nikodym Property*, J. Funct. Anal. 23, 162-176

[16] Campanino M. and Spizzichino F. (1981): *Prediction sufficiency and representation of infinite sequences of exchangeable random variables*, Istituto Matematico "G. Castelnuovo", U. Roma

[17] Choquet G. (1956a): *Existence et unicité des représentations intégrales au moyen des points extrêmaux dans les cônes convexes*, Séminaire Bourbaki, Décembre 1956, 139 (1-15)

[18] Choquet G. (1956b): *Unicité des représentations intégrales au moyen de points extrêmaux dans les cônes convexes réticulés*, C. R. Acad. Sci. Paris Série A, t. 243, 555-557

[19] Choquet G. (1962): *Remarques à propos de le démonstration de l'unicité de P.A. Meyer*, Séminaire Brelot-Choquet-Deny (Theorie de Potentiel) 6, No. 8, 13pp.

[20] Choquet G. (1969): *Lectures in analysis*, Volume 2: *Representation theory*, W.A. Benjamin, Inc.: New York, Amsterdam

[21] Christensen J.P.R. (1983a): *Remarks on Namioka spaces and R.E. Johnson's theorem on the norm separability of the range of certain mappings*, Math. Scand. 52, 112-116

[22] Christensen J.P.R. (1983b): personal communication

[23] Clarkson J.A. (1974): *A characterization of C-spaces*, Ann. of Math. 48, 845-850

[24] Csiszár I. (1984): *Sanov property, generalized I-projection and a conditional limit theorem*, Ann. Probab. 12, No. 3, 768-793

[25] Dalgas K.-P. (1982): *A general extension theorem for group-valued measures*, Math. Nachr. 106, 153-170

[26] Davies E.B. and Vincent-Smith G.F. (1968): *Tensor products, infinite products and projective limits of Choquet simplexes*, Math. Scand. 22, 145-164

[27] Dellacherie C. (1980): *Un cours sur les ensemble analytiques*, in: *Analytic sets*, Instructional Conference ..., London (1978), Academic Press: London, New York

[28] Dinges H. (1970): *Decomposition in Ordered Semigroups*, J. Functional Analysis 5, 436-483

[29] Doob J.L. (1968): *Generalized Sweeping-Out and Probability*, J. Functional Analysis 2, 207-225

[30] Dugundji J. (1968): *Topology*, fourth printing, Allyn and Bacon: Boston

[31] Dynkin E.B. (1971): *Initial and final behaviour of trajectories of Markov processes*, Uspehi Mat. Nauk 26.4, 153-172 (English translation: Russian Math. Surveys 26.4, 165-185)

[32] Dynkin E.B. (1978): *Sufficient Statistics and Extreme Points*, Ann. Probab. 6, 705-730

[33] Edgar G.A. (1975): *A noncompact Choquet theorem*, Proc. Amer. Math. Soc. 49, 354-358

[34] Edgar G.A. (1976): *Extremal Integral Representations*, J. Funct. Anal. 23, 145-161

[35] Edgar G.A. (1977): personal communication

[36] Edgar G.A. (1978): *On the Radon-Nikodym Property and Martingale Convergence*, in: Vector Space Measures and Applications II, Proceedings, Dublin (1977), Lecture Notes in Mathematics 654, 62-76, Springer Verlag: Berlin, Heidelberg, New York

[37] **Edgar G.A. (1983):** personal communication

[38] **Edwards D.A. (1974):** *Suites décroissantes de simplexes*, Séminaire Choquet: Initiation à l'analyse, 13e année, n° 16

[38] **Edwards D.A. (1975):** *Systèmes projectifs d'ensembles convexes compacts*, Bull. Soc. Math. France 103, 225-240

[40] **Eggleston H.G., Grünbaum B. and Klee V. (1964):** *Some semicontinuity theorems for convex polytopes and cell-complexes*, Comm. Math. Helv. 39, 165-188

[41] **Farell R.H. (1962):** *Representation of invariant measures*, Illinois J. Math. 6,3, 447-467

[42] **Föllmer H. (1975):** *Phase transition and Martin boundary*, in: Séminaire de Probabilités IX, Université de Strasbourg, Lecture Notes in Mathematics 465, 305-317, Springer Verlag: Berlin, Heidelberg, New York

[43] **Fremlin D.H., Garling D.J.H. and Haydon R.G. (1972):** *Bounded measures on topological spaces*, Proc. London Math. Soc. (3) 25, 115-136

[44] **Fremlin D. and Pryce I. (1974):** *Semiextremal sets and measure representation*, Proc. London Math. Soc. 28, 502-520

[45] **Fuchssteiner B. and Lusky W. (1981):** *Convex cones*, Mathematics Studies 56, North-Holland Publishing Company

[46] **Gelbaum B.R. and Olmsted J.M.H. (1965):** *Counterexamples in Analysis*, second printing, Holden-Day Inc.: San Francisco, London, Amsterdam

[47] **Georgii H.O. (1979):** *Canonical Gibbs Measures*, Lecture Notes in Mathematics 760, Springer Verlag: Berlin, Heidelberg, New York

[48] **Goldman A. (1977):** *Mesures cylindriques, mesures vectorielles et questions de concentration cylindrique*, Pacific J. Math. 69, No. 2, 385-413

[49] **Graf S. (1980):** *Measurable weak selections*, in: Measure theory Oberwolfach (1979), Proc. of the conf. held at Oberwolfach, Germany, July 1-7, 1979, Lecture Notes in Mathematics 794, 117-140, Springer Verlag: Berlin, Heidelberg, New York

[50] Haydon R. (1976): *An extreme point criterion for separability of a dual Banach space, and a new proof of a theorem of Corson*, Quart. J. Math. Oxford Ser.(2) 27, 379-385

[51] Heyer H. (1982): *Theory of statistical experiments*, Springer Verlag: Berlin, Heidelberg, New York

[52] Hoffmann-Jørgensen J. (1970): *The Theory of Analytic Spaces*, Aarhus: Math. Inst. (Various publication series 10)

[53] Hoffmann-Jørgensen J. (1972): *Weak compactness and tightness of subsets of M(X)*, Math. Scand. 31, 127-150

[54] Hoffmann-Jørgensen J. (1975): *The strong law of large numbers and the central limit theorem in Banach spaces*, in: Various Publ. Ser. No. 24, 74-99, Aarhus Universitet, Matematisk Institut: Aarhus

[55] Hoffmann-Jørgensen J. (1977): *Probability in Banach spaces*, in: Ecole d'Eté de Probabilités de Saint Flour VI-1976, Lecture Notes in Mathematics 598, 1-186, Springer Verlag: Berlin, Heidelberg, New York

[56] Jacod J. and Yor M. (1977): *Études des solutions extrêmales et représentation intégrales des solutions pour certains problèmes de martingales*, Z. Wahrscheinlichkeitstheorie verw. Gebiete 38, 899-912

[57] Jameson G.J.O. (1972): *Convex series*, Proc. Camb. Phil. Soc. 72, 37-47

[58] Jellet F. (1968): *Homomorphisms and inverse limits of Choquet simplexes*, Math. Z. 103, 219-226

[59] Johnson R.E. (1969): *Separate continuity and measurability*, Proc. Amer. Math. Soc. 20, 420-422

[60] Kakutani S. (1943): *Notes on infinite product measures II*, Proc. Japan Acad. 19, 184-188

[61] Kemeny J.G., Snell J.L. and Knapp A.W. (1976): *Denumerable Markov Chains*, second edition, Springer Verlag New York Inc.

[62] Kendall D.G. (1962): *Simplexes and vector lattices*, J. London Math. Soc. (2) 37, 365-371

[63] Kerstan J. (1975): *Ergodische Zerlegungen*, unpublished manuscript

[64] Kerstan J. and Wakolbinger A. (1980): *Opérateurs markoviens et l'ensemble de ses lois de probabilité invariantes*, C. R. Acad. Sci. Paris 291, Sér. A, 163-166

[65] Kerstan J. and Wakolbinger A. (1981): *Ergodic decomposition of probability laws*, Z. Wahrsch. Verw. Gebiete 56, 399-414

[66] Khurana S.S. (1969): *Measures and Barycenters of Measures on Convex Sets in Locally Convex Spaces I*, J. Math. Anal. Appl. 27, 103-115

[67] Kingman J.F.C. (1975a): *Anticipation processes*, in: Perspectives in probability and statistics, Academic Press: London, New York, San Francisco

[68] Kingman J.F.C. (1975b): *Geometrical aspects of the theory of non-homogeneous Markov chains*, Math. Proc. Cambridge Philos. Soc. 77, 171-183

[69] Koch G. and Spizzichino F. eds. (1982): *Exchangeability in Probability and Statistics*, North-Holland: Amsterdam

[70] Köthe G. (1969): *Topological vector spaces I*, Springer Verlag: Berlin, Heidelberg, New York

[71] Kotecký R. and Preiss D. (1979): *Generalized projective limits of measurable spaces and their use for the description of Gibbs states*, Coll. Math. Soc. Janos Bolyai 27, Random Fields, Esztergom (Hungary), 639-645

[72] Kotecký R. and Preiss D. (1980): *Generalized projective limits of measurable spaces and their use for the description of Gibbs states*, unpublished

[73] Kupka J. (1972): *Radon-Nikodym Theorems for Vector Valued Measures*, Trans. Amer. Math. Soc. 169, 197-217

[74] Kuznecov S.E. (1974): *On decomposition of excessive functions*, Soviet Math. Dokl. 15, 121-124

[75] Kuznetsov S.E. (1981): *Any Markov Process in a Borel Space Has a Transition Function*, SIAM, Theory of Probability 25, No. 2, 384-388

[76] Lauritzen S.L. (1982): *Statistical Models as Extremal Families*, Aalborg University Press

[77] Le Cam L. (1975): *Convergence in distribution of stochastic processes*, Univ. of California Publ. in Stat. 2, 207-236

[78] Léger C. and Soury P. (1971): *Le convexe topologique des probabilités sur un espace topologique*, J. Math. Pures. Appl. 50, 363-425

[79] Lindenstrauss J., Olsen G. and Sternfeld Y. (1978): *The Poulsen Simplex*, Ann Inst. Fourier (Grenoble) 28.1, 91-114

[80] Lipecky Z. (1985): personal communication

[81] Mankiewicz P. (1978): *A remark on Edgar's extremal integral representation theorem*, Studia Math. 63, 259-265

[82] Martin-Löf P. (1974): *Repetitive structures*, in: Proceedings of Conference of Foundational Questions in Statistical Inference, Memoirs 1, Barndorff-Nielsen, Blæsild and Schou eds.: Aarhus

[83] Mauldin R.D., Preiss D. and Weizsäcker H.v. (1983): *Orthogonal Transition Kernels*, Ann. Probab. 11, 970-988

[84] Neveu J. (1969): *Mathematische Grundlagen der Wahrscheinlichkeitstheorie*, Oldenbourg Verlag:
München, Wien

[85] Parthasarathy K.R. (1967): *Probability measures on metric spaces*, Academic Press: New York, London

[86] Pestman W.R. (1981): *Convergence of martingales in locally convex Suslin spaces*, thesis, Mathematisch Instituut der Rijksuniversiteit te Groningen

[87] Phelps R.R. (1966): *Lectures on Choquet's theorem*, van Nostrand: New York, Toronto

[88] Poulsen E.T. (1961): *A simplex with dense extreme boundary*, Ann. Inst. Fourier (Grenoble) 11, 83-87

[89] Preston C. (1976): *Random fields*, Lecture Notes in Mathematics 534, Springer Verlag: Berlin, Heidelberg, New York

[90] Robertson A.P. and Robertson W.J. (1967): *Topologische Vektorräume*, Bibliographisches Institut: Mannheim

[91] Rogers C.A. and Shephard, G.C. (1957): *The difference body of a convex body*, Arch. Math. 8, 220-233

[92] Rost H. (1971): *Charakterisierung einer Ordnung von konischen Maßen durch positive L^1-Konstruktionen*, J. Math. Anal. Appl. 33, 35-42

[93] Saint Raymond J. (1974): *Représentation intégrale dans certain convexes*, Sem. Choquet, 14e année, n° 2

[94] Saint Raymond J. (1976): *Boréliens à coupes K_σ*, Bull. Soc. Math. France 104, 389-400

[95] Schaefer H.H. (1971): *Topological Vector Spaces*, Springer Verlag: Berlin, Heidelberg, New York

[96] Scheffer C.L. (1971): *La limite projective des espaces topologiques de Hausdorff compacts dans la catégorie des diffusions continues sous-markoviennes*, C. R. Acad. Sci. Paris 272, 1198-1201

[97] Schwartz L. (1973): *Radon measures on arbitrary topological spaces and cylindrical measures*, Oxford University Press

[98] Schwartz L. (1973a): *Surmartingales régulières à valeurs mesures et désintegrations régulières d'une mesure*, J. d'analyse math. 26, 1-168

[99] Semadeni Z. and Zidenberg H. (1965): *Inductive and Inverse Limits in the Category of Banach Spaces*, Bull. Acad. Pol. Sc. 13, 579-583

[100] Semadeni Z. (1971): *Banach spaces of continuous functions*, PWN-Polish Scientific Publishers: Warszawa

[101] **Sunyach Ch. (1969):** *Une caractérisation des espaces universellement Radon-mesurables*, C. R. Acad. Sc. Paris Série A, t. 268, 864-866

[102] **Talagrand M. (1982):** *Three convex sets*, Preprint

[103] **Thomas E. (1980):** *A converse to Edgar's theorem*, in: Measure theory Oberwolfach (1979), Proc. of the conf. held at Oberwolfach, Germany, July 1-7, 1979, Lecture Notes in Mathematics 794, 497-512, Springer Verlag: Berlin, Heidelberg, New York

[104] **Topsøe F. (1970):** *Topology and measure*, Lecture Notes in Mathematics 133, Springer Verlag: Berlin, Heidelberg, New York

[105] **Topsøe F. (1975):** *Some special results on convergent sequences of Radon measures*, Københavns Universitet, Matematisk Institut, Preprint series No. 3

[106] **Valadier M. (1973):** *Desintegration d'une mesure sur un produit*, C. R. Acad. Sci. Paris Série A, t. 276, 33-35

[107] **Varadarajan V.S. (1963):** *Groups of automorphisms of Borel spaces*, Trans. Amer. Math. Soc. 109,2, 191-220

[108] **Vestergaard J. (~1975):** personal communication

[109] **Wakolbinger A. (1979):** *Singularität von Punktprozessen*, thesis: Linz

[110] **Weizsäcker H.v. (1975):** *Der Satz von Choquet-Bishop-de Leeuw für konvexe nicht-kompakte Mengen straffer Maße über beliebigen Grundräumen*,, Math. Z. 142, 161-165

[111] **Weizsäcker H.v. (1976):** *A note on infinite dimensional convex sets*, Math. Scand. 38, 321-324

[112] **Weizsäcker H.v. (1977):** *Einige maßtheoretische Formen der Sätze von Krein-Milman und Choquet*, Habilitationsschrift: München

[113] **Weizsäcker H.v. (1978):** *Strong measurability, liftings and the Choquet-Edgar theorem*, in: Vector Space Measures and Applications II, Proceedings, Dublin (1977), Lecture Notes in Mathematics 654, 209-218, Springer Verlag: Berlin, Heidelberg, New York

[114] Weizsäcker H.v. (1982): personal communication

[115] Weizsäcker H.v. and Winkler G. (1979): *Integral Representation in the Set of Solutions of a Generalized Moment Problem*, Math. Ann. 246, 23-32

[116] Weizsäcker H.v. and Winkler G. (1980): *Non-compact extremal integral representations: some probabilistic aspects*, in: Functional Analysis: Surveys and Recent Results II, Mathematics Studies 38, 115-148, North-Holland: Amsterdam, Oxford, New York

[117] Winkler G. (1977): *Über die Integraldarstellung in konvexen nicht kompakten Mengen straffer Maße*, Thesis: München

[118] Winkler G. (1978): *On the integral representation in convex non-compact sets of tight measures*, Math. Z. 158, 71-77

[119] Winkler G. (1980): *A Choquet ordering and unique decompositions in convex sets of tight measures*, Bull. Greek Math. Soc. 21, 32-46

[120] Winkler G. (1983): *Inverse limits of simplices and applications in stochastics*, Habilitationsschrift, Universität München

[121] Winkler G. (1984a): *A note on the extension of weak Radon probability measures on locally convex spaces to strong Radon probability measures*, Supplemento ai Rendiconti del Circolo Matematico di Palermo, Serie II, nu. 3, 381-384

[122] Winkler G. (1984b): *Inverse limits need not exist in the category of compact spaces and Feller kernels: a counterexample*, Supplemento ai Rendiconti del Circolo Matematico di Palermo, Serie II, nu. 5, 155-159

[123] Yor M. (1978): *Quelques résultats sur certaines mesures extrêmales. Application à représentation des martingales*, in: Measure Theory, Applications to Stochastic Analysis, Proceedings, Oberwolfach Conference, Germany, July 3-9, 1977, 27-36, Lecture Notes in Mathematics 695, Springer Verlag: Berlin, Heidelberg, New York

INDEX OF SYMBOLS

SUBJECT INDEX